Arduino iOS Blueprints

Integrate the Arduino and iOS platforms to design amazing real-world projects that sense and control external devices

Fabrizio Boco

BIRMINGHAM - MUMBAI

Arduino iOS Blueprints

First published: September 2015

Production reference: 1240915

Published by Packt Publishing Ltd.
Livery Place
35 Livery Street
Birmingham B3 2PB, UK.

ISBN 978-1-78528-366-6

www.packtpub.com

Credits

About the Author

Fabrizio Boco was born in Italy in 1964. He started with electronics when he was a teen and with programming in 1980. In 1992, he received his degree in electronics engineering. Currently, he is a freelance IT manager and architect and has more than 20 years of experience in consulting for private and public companies in Italy. Even when he occupied executive positions, he worked on the design and implementation of IT projects (mostly on enterprise applications, data warehousing, and business intelligence). He has been an iOS developer since 2009, and he has designed and developed Arduino Manager, which is an iOS, Mac OS X, and Android application that can be used to control Arduino (and Arduino-compatible) boards through a powerful and easy-to-use interface, that is based on more than 20 specialized widgets (a switch, knob, display, and gauge, among others). Fabrizio believes that his engineering skills are hardcoded within his DNA; the studies and experiences only completed them.

About the Reviewers

Tomi Dufva completed an MA in fine arts. He is a doctoral researcher at Aalto University and the cofounder of Art and Craft School Robotti (for more information, visit www.kasityokoulurobotti.fi). He lives and works in Turku as a visual artist, art teacher, and researcher. He researches creative coding at Aalto University, School of Arts, Design, and Architecture. Tomi specializes in code literacy, maker culture, pedagogical use of code, and the integration of painting and drawing with electronics and code. Tomi has taught in various schools, from kindergartens to universities. You can go through Tomi's research by visiting his blog (www.thispagehassomeissues.com).

Heqing Yan is a second year mechatronics engineering student at the University of Waterloo. He is currently studying for a bachelor's degree in applied science. He worked as a software engineer for Pearl App, Inc. and developed an iOS app called *Pearl21*. To know more about him, visit his website (http://www.scottie.me/).

www.PacktPub.com

Support files, eBooks, discount offers, and more

For support files and downloads related to your book, please visit www.PacktPub.com.

Did you know that Packt offers eBook versions of every book published, with PDF and ePub files available? You can upgrade to the eBook version at www.PacktPub.com and as a print book customer, you are entitled to a discount on the eBook copy. Get in touch with us at service@packtpub.com for more details.

At www.PacktPub.com, you can also read a collection of free technical articles, sign up for a range of free newsletters and receive exclusive discounts and offers on Packt books and eBooks.

https://www2.packtpub.com/books/subscription/packtlib

Do you need instant solutions to your IT questions? PacktLib is Packt's online digital book library. Here, you can search, access, and read Packt's entire library of books.

Why subscribe?

- Fully searchable across every book published by Packt
- Copy and paste, print, and bookmark content
- On demand and accessible via a web browser

Free access for Packt account holders

If you have an account with Packt at www.PacktPub.com, you can use this to access PacktLib today and view 9 entirely free books. Simply use your login credentials for immediate access.

This book is dedicated to

my father, who taught me the basics,

my mother, who is still supporting me,

my wife, Laura, who is sharing with me the ups and downs of life,

my daughter, Camilla, who helped me with the English,

but mostly to all the people who tried to make my life harder!

Table of Contents

Preface   **v**

Chapter 1: Arduino and iOS – Platforms and Integration   **1**

Hardware and software requirements   **2**

Hardware requirements for the Arduino platform   2

Software requirements for the Arduino platform   3

Hardware requirements for the iOS platform   4

Software requirements for the iOS platform   4

Arduino and the development environment setup   **4**

IDE installation   5

iOS and the development environment setup   **6**

Xcode installation   6

Communication methods between Arduino and iOS devices   **7**

TCP/IP versus Bluetooth   8

Summary   **9**

Chapter 2: Bluetooth Pet Door Locker   **11**

Door locker requirements   **12**

Hardware   **12**

Required materials and electronics components   12

Assembly latch and servo motor   13

Electronic circuit   14

Arduino code   **17**

Installing additional required libraries   19

Initializing global variables and libraries   20

Setup code   20

Main program   22

Testing and tuning the Arduino side   27

iOS code	**28**
Creating the Xcode project	28
Designing the application user interface for BLEConnectionViewController	31
Designing the application user interface for PetDoorLockerViewController	39
Writing code for BLEConnectionViewController	41
Writing code for PetDoorLockerViewController	46
Testing the iOS app	53
How to go further	**53**
Different types of sensors	**54**
Summary	**55**
Chapter 3: Wi-Fi Power Plug	**57**
Wi-Fi power plug requirements	**58**
Hardware	**58**
Additional electronics components	58
Electronic circuit	58
Arduino code	**61**
Setup code	62
Main program	62
iOS code	**66**
Creating the Xcode project	66
Adding a new view controller	68
Adding a class for storing the information of each activation	73
Designing the application user interface for WiFiConnectionViewController	75
Designing the application user interface for PowerPlugViewController	76
Designing the application user interface for ActivationsTableViewController	78
Writing code for the WiFiConnectionViewController	85
Writing code for AppDelegate	88
Writing code for PowerPlugViewController	94
Writing code for ActivationsTableViewController	95
Writing code for ActivationTableViewController	99
Testing and tuning	99
How to access the power plug from anywhere in the world	**100**
Port forwarding	101
Dynamic DNS	102
How to go further	**103**
Summary	**103**

Chapter 4: iOS Guided Rover   105

iOS guided rover requirements   106
Hardware   106
Additional electronic components   106
What's an accelerometer?   107
Electronic circuit   108
How to make the rover turn   112
How to mount the accelerometer   112
Arduino code   113
Setup code   114
Motor control functions   114
Main program   116
iOS code   120
Creating the Xcode project   120
Writing code for BLEConnectionViewController   127
Writing code for RoverViewController   127
Code to control the rover manually   130
Testing the Rover with manual driving   132
Code for controlling the rover by the means of the iOS accelerometer   133
Driving the rover by the means of the iOS device movement   136
Code for controlling the rover by voice commands   137
Driving the rover by voice commands   141
Testing and tuning   141
How to go further   142
Summary   143

Chapter 5: TV Set Constant Volume Controller   145

Constant Volume Controller requirements   145
Hardware   146
Additional electronic components   146
Electronic circuit   147
Arduino code   148
Decoder setup code   149
Decoder main program   149
Setup code   150
Main program   150

iOS code — **151**
Creating the Xcode project — 151
Designing the user interface for VolumeControllerViewController — 153
Writing code for BLEConnectionViewController — 154
Writing code for VolumeControllerViewController — 154
Testing and tuning — 161
How to go further — **162**
Summary — **162**

Chapter 6: Automatic Garage Door Opener — **163**
iBeacon – a technical overview — **164**
The garage door opener requirements and design constraints — **166**
Hardware — **170**
Additional electronic components — 170
Electronic circuit — 171
Arduino code — **175**
Setup code — 176
Main program — 177
iOS code — **178**
Creating the Xcode project — 178
Designing the user interface for BLEConnectionViewController — 180
Designing the user interface for GarageViewController — 183
Designing the user interface for PinsViewController — 185
Writing code for BLEConnectionViewController — 187
Writing code for GarageViewController — 191
Writing code for PinsViewController — 204
Testing and tuning — 205
How to go further — **208**
Summary — **209**
Index — **211**

Preface

It was the fall of 2011, and I was working on some iOS apps. On my desk lay an Arduino board that was almost unused. A thought came to me—it would be great to integrate the iOS platform and Arduino. I could control almost everything from anywhere, from my home to industrial machines.

I started working on this and I eventually designed Arduino Manager. It is a general-purpose iOS application based on widgets, which acquires data from Arduino and depicts it with gauges, graphs, and other means. Arduino Manager allows you to control an Arduino board through switches, knobs, sliders, and other means (more information on this is available at `http://apple.co/1NPfL6i`).

This book shows some of the basic techniques that I developed during these years to make the Arduino and iOS devices work together.

I will show you how to build five amazing projects, and you will learn how the additional electronics around Arduino work. You will also learn how to use digital and analog sensors, program the Arduino board, and develop iOS applications that can transfer data with Arduino. Projects are described in detail, providing you with a learning tool and not just some sketches or some iOS code to copy.

I promise that you will not get bored.

What this book covers

Chapter 1, *Arduino and iOS – Platforms and Integration*, will give you a brief introduction to the two platforms and introduce the integration methods between them. Moreover, you will learn how to set up the development environments on both the platforms and get ready for the projects in the subsequent chapters.

Chapter 2, Bluetooth Pet Door Locker, develops a project that helps you automatically lock the pet door at night by measuring the external light, monitoring whether it is locked or unlocked, and manually operating it as needed. This is done by using an iOS device.

Chapter 3, Wi-Fi Power Plug, is about learning how to make a smart power plug that is controlled thorough Wi-Fi. This is not based on the traditional relay technique, and the iOS application is full of useful features.

Chapter 4, iOS Guided Rover, will teach you how to control a rover robot by using voice commands and moving your iOS device.

Chapter 5, TV Set Constant Volume Controller, is for people who are bored by the high volume of commercials. You will learn how to make an automatic IR controller that keeps the TV set's volume constant.

Chapter 6, Automatic Garage Door Opener, covers a project that will allow you to automatically open your garage door by just getting close enough to it, without even touching the iOS device. The project takes advantage of the iBeacon technology.

What you need for this book

To build the hardware side of the projects, you will need an Arduino board and some other electronic components.

To develop the Arduino firmware, you just need the Arduino IDE, which can be downloaded for free from the Arduino site.

To develop the iOS applications, you need the Xcode development environment (available for free from Apple). Since the Xcode is available only for the MAC platform, you need an Intel-based MAC computer. Most of the projects are based on Bluetooth 4 protocol. Hence, you need an iOS device that supports this protocol.

Everything is explained in detail in *Chapter 1, Arduino and iOS – Platforms and Integration*.

Who this book is for

This book is a technical guide for Arduino and iOS developers who have a basic knowledge of the two platforms but want to learn how to integrate them. The book is complete with a lot of external references to additional documentation and learning materials. Therefore, even a less experienced reader can improve his knowledge on the covered subjects.

Conventions

In this book, you will find a number of text styles that distinguish between different kinds of information. Here are some examples of these styles and an explanation of their meaning.

Code words in text, database table names, folder names, filenames, file extensions, pathnames, dummy URLs, user input, and Twitter handles are shown as follows: " Usually, the browser will download the file in your `Downloads` folder."

A block of code is set as follows:

```
if (deviceIdentifier!=nil) {

        NSArray *devices = [_centralManager
retrievePeripheralsWithIdentifiers:@[[CBUUID UUIDWithString:deviceIde
ntifier]]];
        _arduinoDevice = devices[0];
        _arduinoDevice.delegate = self;
    }
```

When we wish to draw your attention to a particular part of a code block, the relevant lines or items are set in bold:

```
if (deviceIdentifier!=nil) {

        NSArray *devices = [_centralManager
retrievePeripheralsWithIdentifiers:@[[CBUUID UUIDWithString:deviceIde
ntifier]]];
        _arduinoDevice = devices[0];
        _arduinoDevice.delegate = self;
    }
```

New terms and **important words** are shown in bold. Words that you see on the screen, for example, in menus or dialog boxes, appear in the text like this: "Alternatively, you can run it from Launchpad or by navigating to **Finder | Applications | App Store**."

Warnings or important notes appear in a box like this.

Tips and tricks appear like this.

Reader feedback

Feedback from our readers is always welcome. Let us know what you think about this book—what you liked or disliked. Reader feedback is important for us as it helps us develop titles that you will really get the most out of.

To send us general feedback, simply e-mail feedback@packtpub.com, and mention the book's title in the subject of your message.

If there is a topic that you have expertise in and you are interested in either writing or contributing to a book, see our author guide at www.packtpub.com/authors.

Customer support

Now that you are the proud owner of a Packt book, we have a number of things to help you to get the most from your purchase.

Downloading the example code

You can download the example code files from your account at http://www.packtpub.com for all the Packt Publishing books you have purchased. If you purchased this book elsewhere, you can visit http://www.packtpub.com/support and register to have the files e-mailed directly to you.

Errata

Although we have taken every care to ensure the accuracy of our content, mistakes do happen. If you find a mistake in one of our books—maybe a mistake in the text or the code—we would be grateful if you could report this to us. By doing so, you can save other readers from frustration and help us improve subsequent versions of this book. If you find any errata, please report them by visiting http://www.packtpub.com/submit-errata, selecting your book, clicking on the **Errata Submission Form** link, and entering the details of your errata. Once your errata are verified, your submission will be accepted and the errata will be uploaded to our website or added to any list of existing errata under the Errata section of that title.

To view the previously submitted errata, go to https://www.packtpub.com/books/content/support and enter the name of the book in the search field. The required information will appear under the **Errata** section.

Piracy

Piracy of copyrighted material on the Internet is an ongoing problem across all media. At Packt, we take the protection of our copyright and licenses very seriously. If you come across any illegal copies of our works in any form on the Internet, please provide us with the location address or website name immediately so that we can pursue a remedy.

Please contact us at `copyright@packtpub.com` with a link to the suspected pirated material.

We appreciate your help in protecting our authors and our ability to bring you valuable content.

Questions

If you have a problem with any aspect of this book, you can contact us at `questions@packtpub.com`, and we will do our best to address the problem.

1
Arduino and iOS – Platforms and Integration

This chapter will give you a brief introduction to the Arduino and iOS platforms and the integration methods between them. Moreover, you will learn how to set up the development environments on both the platforms and get them ready for the projects in the following chapters.

We will assume that you already have some basic knowledge of the two platforms and electronics, and you are able to build a circuit, at least by using a breadboard. However, the main subject of the chapter is to learn how to integrate the two platforms together.

Arduino has been an Open Hardware device since its early origin, and you can easily find any kind of information you need about it. Conversely, the iOS platform is not very open, especially from the hardware point of view. You cannot design and build a hardware device that works with iOS devices without joining an Apple dedicated program (MFi). The program has strong requirements that only large companies can fulfill.

More information on Arduino and iOS
More information on Arduino and iOS development can be found at `http://www.arduino.cc` and `http://apple.co/1HThS1O` respectively.

Nevertheless, at the end of the chapter, you will learn how to transfer data between the two platforms in ways that also allow your applications to be sold on the iTunes App Store. This consists of nothing that is too complex. We are going to use TCP/IP or Bluetooth BLE.

The following are the topics that will be covered in this chapter:

- Hardware and software requirements
- Arduino and the development environment setup
- iOS and the development environment setup
- Communication methods between the Arduino and iOS devices

Hardware and software requirements

To implement all the projects in this book, you will need some hardware and software components, which can be easily bought from any of your local stores or over the Internet.

Hardware requirements for the Arduino platform

To execute the projects in this book, you mainly need Arduino UNO and the following additional hardware components:

1. Arduino UNO R3 (`http://bit.ly/1IInOke`).
2. A USB cable (A to B type).
3. A 9V external DC power supply (optional but recommended).
4. An official Wi-Fi Shield (`http://bit.ly/1UQgq9v`, `http://bit.ly/1i5k1Cn`).
5. A Bluetooth BLE nRF8001 breakout board from Adafruit (`http://bit.ly/1MvkyJm`).
6. A digital multimeter (optional but strongly recommended).
7. A breadboard (the larger, the better).
8. Some tools (skewdrivers, pliers, tweezers, nippers, and so on).
9. Some Male/Male and Male/Female jumper wires (the more, the better; they are never enough!)
10. Some electronics components, which will be shown chapter by chapter.
11. A rover robot for a project in *Chapter 5*, *TV Set Constant Volume Controller*.
12. An iBeacon for a project in *Chapter 6*, *Automatic Garage Door Opener* (`http://apple.co/1GXnt7Z`).

Buy original products only!

There are a lot of counterfeited Arduino products, especially online, that are sold for a few dollars. Apart from the moral considerations, an original product will provide you with much more quality and certainty to right functioning. Projects in this book are developed and tested on original products. You have been warned!

The official Wi-Fi Shield

The official Wi-Fi Shield is not cheap, and it can be tough to make it work with iOS sometimes. Many other products are cheaper. Some are compatible, others are not. If you choose to use a different Wi-Fi Shield, you need to be sure that it works with the Wi-Fi library that is included in the Arduino IDE. Otherwise, you will have to modify the Arduino code in this book on your own. You can also buy this Shield at http://bit.ly/1i5k1Cn.

Arduino.cc versus Arduino.org conflict

At the time of writing this book, Arduino split in two companies — arduino.cc and arduino.org. They are in the middle of a legal battle, and there is a lot of confusion about products with the same name, which are sold by both the companies. Though the situation is getting clearer in USA and Asia, it is still unpredictable in Europe. In this book, we will be using the products and development environment from arduino.cc.

The Bluetooth Breakout board

The Bluetooth device that we chose has been tested many times on different projects and works well. There are many other devices that are available on the market and which can do exactly the same job. If you decide to use a different Bluetooth device, be sure that is a BLE (Bluetooth Low Energy) device, which is also known as Bluetooth 4.0, Bluetooth Low Energy and Bluetooth Smart. You can easily write iOS applications, which communicate through Bluetooth only if it is Bluetooth BLE.

Software requirements for the Arduino platform

All that you need is Arduino IDE 1.6.4, the libraries included in the IDE, and some additional libraries that are available online. You will be instructed to download them when needed.

Hardware requirements for the iOS platform

The Apple development environment that is needed to write iOS applications (Xcode) is available only for the Mac OS X operating system (the 10.10.3 Yosemite version). So, you need a recent Intel-based Mac computer to run Yosemite.

Moreover, you need an iOS device (iPhone, iPad, or iPod touch) that runs iOS 8.4 and supports Bluetooth 4.0.

If you are already a member of the **Apple Developer Program** (**ADP**) for iOS (which costs about $99 per year), your life will be easy. Unfortunately, Apple doesn't allow you to upload programs that are written with Xcode to your own device if you are not a member of ADP. You can still run the programs in the device simulator, but it will not be able to simulate the Bluetooth BLE subsystem. This will be a problem for projects that use Bluetooth BLE. Anyway, adapting the code presented in this book to use Wi-Fi instead of Bluetooth BLE should not be too complex.

Xcode and the new Apple rules

At the time of writing this book, Apple announced a new version of iOS (version 9) and Xcode (version 7) that are now in their beta versions. Some restriction rules will be removed from these versions, and the user should be able to upload self-written apps to their own devices even if they are not subscribed to the Apple Developer Program.

Software requirements for the iOS platform

As already mentioned, to write applications for the iOS platform, you need Xcode (6.4 at the time of writing this book) and some additional open source libraries that are available online. You will be instructed to download them when needed. Xcode runs exclusively on Mac with OS X (version 10.10.3 Yosemite or higher). Even if most of the code of this book can run on the previous versions of iOS, your iOS device has to be able to run iOS 8.3 and higher since it is the version that is used across the book.

Arduino and the development environment setup

Arduino is a very basic but powerful microprocessor board. It allows you to acquire digital and analog inputs, process them, and control external devices.

You can find all the details about this device at http://bit.ly/1IInOke.

What makes it different from many other similar (or even more powerful) devices on the market is that it has a very simple development environment (IDE) and plenty of libraries, which allow you to make projects in a matter of hours. Most of these libraries are already available in the IDE or are very easy to add.

Arduino UNO – Arduino MEGA

In this book, we are going to refer to Arduino UNO. Many other boards from Arduino are available with different hardware characteristics, FLASH, and the RAM memory size. Most of the projects in this book will run on Arduino MEGA with no changes in their code and minor changes in hardware.

Other platforms

Every few weeks, a new platform that is compatible with Arduino comes out. Two of these platforms, which are known to be compatible with the projects in this book, are RFduino, which supports Bluetooth BLE without any additional hardware, (for more information, visit http://www.rfduino.com) and Teensy (to know more about Teensy, visit https://www.pjrc.com/teensy). Some changes in hardware and/or software may be necessary to work with these platforms.

IDE installation

The process of installing the IDE is very quick and easy, especially because you don't need to install additional drivers on OS X.

You can use the following steps to have the development environment ready in minutes:

1. Make sure that you have installed an application that can open a ZIP file. If you don't have such an application, you can find them available for free on the App Store (The Unarchiver may be an option, which can be downloaded from http://apple.co/1gT7W2D).

2. Download Arduino IDE 1.6.4 from http://bit.ly/1gT8u8E.

3. Usually, the browser will download the file in your Downloads folder. Open this folder and unzip the ZIP file (Arduino-1.6.4-macosx.zip).

4. Move the Arduino.app file to your app folder.

The development environment is now ready to run.

Just to be sure that everything is working as expected, upload one of the included examples (Blink is fine) to your UNO. Refer to http://bit.ly/1KsUhqv in order to properly configure the board and port in the IDE.

The IDE version to use

The code in this book was tested with IDE 1.6.4, which is the latest version that is available at the time of writing this book. Even if a new version of the IDE is available, you should use the suggested version. The libraries that are included in different versions of the IDE usually vary, and this may cause unexpected behavior. You can test a newer IDE once you have completely built and tested the project on Arduino and iOS.

iOS and the development environment setup

In the last few years, commonly named smartphones have become the most used personal devices. iPhone is one of the most used smartphones all over the world, and iPads have replaced personal computers in many cases.

iPhone, iPad, and iPod touch run the same operating system, which is called iOS. They have a neat and homogeneous user interface, which allows its users to interact with the device through simple gestures (such as tapping and double-tapping the screen) or complex gestures (such as swiping or pinching the screen).

Applications that run on both iPhone and iPad can be easily written with minor changes, and the platforms are so powerful that almost any application can run on them.

We can directly go to the process of installing the development environment so that we can start having fun with iOS and Arduino.

Xcode installation

You can perform the following steps to install Xcode in minutes:

1. Open the App Store application. Usually, you will find it in the Dock. Alternatively, you can run it from Launchpad or by navigating to **Finder | Applications | App Store**.

2. Connect to the App Store by using your account (browse **Store | Sign in**).

3. Locate Xcode (navigate to **Categories | Developer Tools** or search for it).

4. Click on the **Get** button to start downloading.

In a short span of time, the application will be downloaded and installed (depending on the speed of your Internet connection).

Communication methods between Arduino and iOS devices

iOS is not so open platform, especially from the hardware point of view. Apple requires that the hardware devices connected to iPhone or iPad should not only be compliant with straight requirements, but also be certified by Apple itself. For this, they need to join an Apple dedicated program (MFi: http://apple.co/1PwSeWO). The requirements to join the program are very tight, and only a large company can fulfill them.

Nevertheless, Apple allows the use of TCP/IP and Bluetooth BLE to communicate with any external device. These are the two methods that we are going to use all over the book to transfer data with Arduino. On Arduino's side, we have to choose additional hardware that is compatible with these two methods or, in a more technical jargon, these two protocols.

Arduino provides the following two Shields for TCP/IP communication:

- The Ethernet Shield
- The Wi-Fi Shield

Both these Shields include an SD card bay to store data.

The Ethernet Shield is relatively cheap and very reliable. Since it uses an Ethernet connection, it's drawback is that it needs physical wiring to your home network or router. This reduces the flexibility of your projects.

Conversely, the Wi-Fi Shield allows you to install your Arduino board everywhere in your house without the awkwardness of wires. This is the reason why we chose to use the Wi-Fi Shield in this book.

The other connection method (protocol) that is allowed by Apple to transfer data to and from external devices is Bluetooth BLE (also known as Bluetooth 4.0). This protocol consumes less energy, but it's not compatible with the previous versions. If your iOS device is quite new, Bluetooth BLE will be supported. Please check this out on your device page at the Apple site.

BLE supporting devices

If you wish to check whether a device supports BLE, visit `http://bit.ly/1blI106`.

Arduino doesn't provide a Bluetooth BLE Shield, but other vendors do. We chose the Bluetooth BLE nRF8001 breakout board from Adafruit.

Bluetooth 4.0

You can learn more about Bluetooth 4.0 by visiting `http://bit.ly/1Pj9caw`.

TCP/IP versus Bluetooth

You may wonder which is the best communication method for iOS and Arduino. There is no right answer. This actually depends on your project and your requirements. The following table shows the main pros and cons of each method:

Method	Pros	Cons
Wi-Fi	• Arduino can be located almost everywhere in your house. • The iOS device can be everywhere (if the network is properly set, even on the other side of the world). • An SD card can be used to store data.	• This is expensive. • Even if data could be transferred with up to 54 Mbps, the Wi-Fi Shield is not so responsive in transferring data. • The power consumption is so high that you cannot use a battery to power Arduino and the Wi-Fi Shield.

Method	Pros	Cons
Bluetooth	• Arduino can be located almost everywhere in your house. • BLE devices consume much less power. So, a battery can be used to power it. • Data transfer speeds reach up to 1 Mbps but with no latency.	• The iOS device has to be near Arduino (about 100 meters; lesser for indoors). • No SD card is available on the board if you wish to store additional data.

Summary

In this chapter, you mainly learned the basics of integrating the Arduino and iOS devices. In the following chapters, you will learn how to write code for this integration in practice.

Moreover, you installed the Arduino IDE to write and upload programs to Arduino and Xcode to write and upload programs to your iOS devices.

So, sit back. We just got the ball rolling! Let's get rolling!

2
Bluetooth Pet Door Locker

This project is about a pet door controller that, by measuring external light and temperature, locks or unlocks your house's pet door. Through an iOS device you can check the status of the door (locked or unlocked) and overrun the logic implemented on Arduino for manually locking the door.

In this project, you will learn how to work with analog sensors, switches, 1-Wire sensors (to measure temperatures) and control a servo motor wired to Arduino. Moreover, you will wire a Bluetooth 4.0 board to Arduino to communicate with your iOS device.

Then, you will learn how to write a Bluetooth iOS application to send and receive data from Arduino.

Towards the end of this chapter, we will discuss different types of sensors and their communication protocols with Arduino, in order to learn how to manage the most used technologies to measure quantities.

This project requires some DIY skills in order to mount the locker to the pet door and to link it to the servo motor.

The chapter is organized into the following sections:

- **Door locker requirements**: We will briefly recap the project requirements
- **Hardware**: We will describe the hardware and the electronic circuit needed for the project
- **Arduino code**: We will write the code for Arduino to control the latch and communicate with the iOS device
- **iOS code**: We will write the code for the iOS device

- **How to go further**: More ideas to improve the project and learn more
- **Different types of sensors**: A quick overview of analog and digital sensors, low-level communication protocols, and their pros and cons

Door locker requirements

We are going to build an automatic system that allows you to lock the pet door in the following different scenarios:

- At night, when the outside light is less luminous than a predefined threshold
- When the external temperature is too low or too high for your beloved pet, that is, when the temperature is below a predefined threshold or above a predefined threshold

Moreover, we need to see the external temperature on our iOS device, we also need to know if the pet door is locked or unlocked, and override the automatic behavior, manually locking the door.

Hardware

In this project, we have to build an electro-mechanical device, based on a latch and a servo motor to lock/unlock the pet door. Moreover, we need to build a simple electronic circuit to read sensors.

Required materials and electronics components

To build the locker you need little hardware that will be available at your local hardware store:

- A small door latch, which is best if flat and without any notch. Since it has to be operated by a servo motor it has to slide very smoothly.
- Some nuts and bolts to mount the latch to the door.
- A few centimeters of metal wire (diameter of 2 mm or so).

Other required components are:

- Mini servo motor powered at 5V.
- Normally closed magnetic switch and a small magnet.
- A photoresistor.
- Resistors: 2 × 10K, 1 × 4.7K.
- A digital DS18B20 temperature sensor. It's more expensive than analog sensors (like TMP 35), but it's much more precise and the readings are almost insensitive to the voltage fluctuations and electric noise.
- Adafruit Bluefruit LE nRF8001 breakout (`http://www.adafruit.com/product/1697`).

Assembly latch and servo motor

To assemble the latch, you can refer to the following picture.

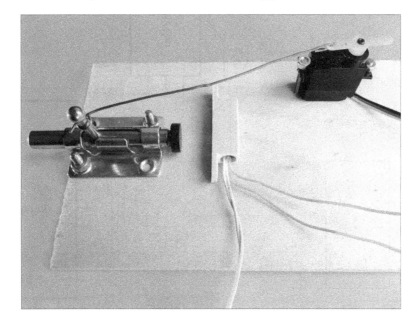

The small magnet is glued on the latch, so that when it is completely retracted, the magnetic switch is open.

The photoresistor and the temperature sensor have to be placed outside the house and wired to Arduino. They need to be protected from humidity and dust, so it's better to set them inside a small plastic box. The box has to be drilled so that the air can freely circulate to help the temperature sensor to measure correctly. Moreover, the photoresistor should not be exposed to direct light to avoid blinding. A small plastic tube can do the job.

Electronic circuit

The following picture shows the electric diagram of the electronic circuit that we need for the project:

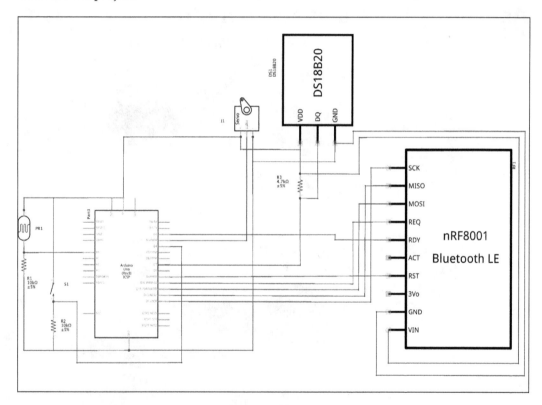

The following picture shows how to mount the circuit on a breadboard:

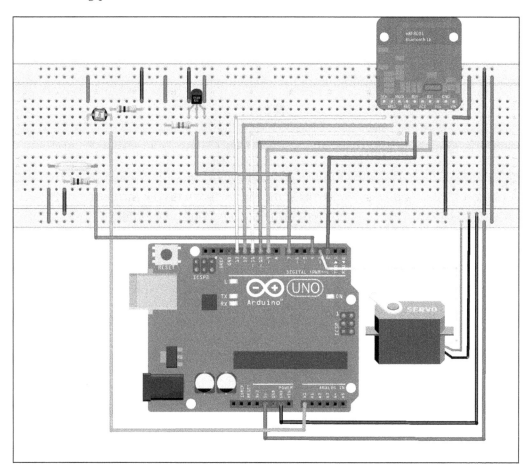

For quick reference, the following list recaps the **nRF8001** wiring:

- SCK to digital pin 13
- MISO to digital pin 12
- MOSI to digital pin 11
- REQ to digital pin 10
- RST to digital pin 9
- RDY to digital pin 2

If you are using **Arduino MEGA** instead of **UNO**, the wiring of **nRF8001** board has to be changed as follows:

- SCK to digital pin 52
- MISO to digital pin 50
- MOSI to digital pin 51
- REQ to digital pin 10
- RST to digital pin 9
- RDY to digital pin 2

Light is measured using the photoresistor and an analog input of Arduino (A0). A photoresistor is a device whose resistance decreases with the increasing of the incident light intensity. In the circuit, the photoresistor is in a voltage divider with R1. Voltage across R1 increases if the light on the photoresistor increases, and it's measured using the analog pin A0 of Arduino.

R1 forces to ground the Arduino input when the photoresistor is not lighted and its resistance is very high.

In the Arduino code, you can get the value of the voltage at an analog pin using the `analogRead` function. It returns a value in the range 0-1023 which is proportional to the voltage applied to the analog pin.

The latch position is determined using a magnetic switch and a digital input of Arduino (D4). The magnetic switch closes when the magnet on the latch is close to it. The switch is in a voltage divider with R2. The voltage across R2 is almost 5V when the switch is closed, or about 0V when the switch is open.

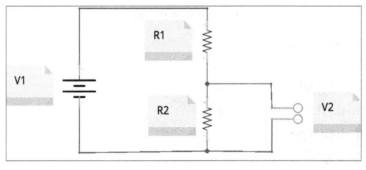

Voltage divider

The voltage divider is a simple circuit made of two resistors. Assuming that the current that flows outside the circuit is negligible, V2 is approximately *V1*R2/(R1+R2)*.

Using an Arduino digital input you can determine if the switch is open or closed.

 In the Arduino code, you can get the value of a digital pin using the `digitalRead` function.

The temperature sensor is a digital sensor that uses the 1-Wire protocol, and it is read using a digital pin.

 To read the value of the temperature sensor you have to use the `OneWire` library and the `DallasTemperature` library through which a function directly returns the temperature in Celsius.

The servo motor requires a PWM signal to be controlled. Pulse width modulation is a technique for getting analog results with digital means. It is a square wave with a period of 2 ms. The width of the positive pulse determines the rotation of the servo (for example, a 1.5 ms pulse will make the motor turn to the 90-degree position). Fortunately, the servo library (already included in the IDE) hides the complexity and we can move the motor by just calling a function with the desired motor position.

Arduino code

The full code of this project can be downloaded from here:

```
https://www.packtpub.com/books/content/support
```

For a better understanding of the explanations in the next paragraphs, you should open the downloaded code while reading.

Each Arduino program almost always has the following structure.

```
#include <library_1.h>
#include <library_2.h>

#define SOMETHING A_VALUE
```

```
// Function prototypes

void callback_1();


// Global variables

boolean  var_1;


// Called only once at power on or reset

void setup() {

...
}

// Called over and over again

void loop() {
...
}

// Callbacks

void callback_1() {
...
}

// Additional functions

void function_1() {
...
}
```

The instruction #include <library_1.h> tells the compiler to use a library and the instruction #define SOMETHING A_VALUE tells the compiler to replace SOMETHING with A_VALUE all over the program. It's easier to make changes if you use defined values. If you need to replace A_VALUE with something else, you can do it at a single place instead of in all instances in the code.

Function prototypes are generally used for functions, which are used during the code but their bodies are put at the end of the program source. Global variables are variables whose values have to be preserved across the execution of the program. The function `setup` is used for initializing libraries and variables; it's called only once, at the board's power up or reset. The function `loop` instead is called over and over again. Variables defined in the loop function lose their value across calls; this is why we need global variables. Callback functions are called from libraries when an event happens, or data are available for processing. Instead, functions that are called from the loop help to make the code easier to read, maintain, and debug.

Installing additional required libraries

For this project we need some libraries. Some are already available in the Arduino IDE and others have to be added (for example, `OneWire`, `DallasTemperature`, and `Adafruit_BLE_UART`).

To add them, follow this simple procedure:

1. Select the menu item **Sketch | Include Library | Manage Libraries**.
2. In the search field enter `OneWire`.
3. Select the row with **OneWire** and click **Install** (see picture below).
4. Enter `DallasTemp` in the search field, click on **MAX31850** and then **Install**.
5. Enter `nRF8001` in the search field, click **Adafruit nRF8001** and then **Install**.

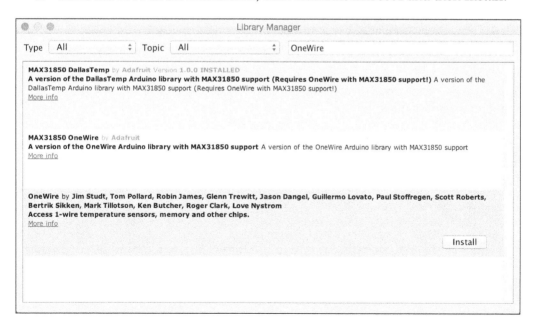

Initializing global variables and libraries

The temperature sensor uses the 1-Wire protocol for communicating with Arduino, so we need the `OneWire` library. Fortunately another library (`DallasTemperature`) is available to make temperature readings very easy. We need to create global variables for the two libraries:

```
OneWire oneWire(ONE_WIRE_BUS);
DallasTemperature sensors(&oneWire);
```

And one for storing the sensor address:

```
DeviceAddress temperatureSensorAddress;
```

The global variables for the nRF8001 library are in a line:

```
Adafruit_BLE_UART uart = Adafruit_BLE_UART(ADAFRUITBLE_REQ,
ADAFRUITBLE_RDY, ADAFRUITBLE_RST);
```

 If you change the pins used in the previous line, you have to also change them in the circuit wiring.

To control the servo motor we need another global variable:

```
Servo myservo;
```

We also need two more global Boolean variables:

```
boolean   iOSConnected;
boolean   manuallyLocked;
```

The first is true when the iOS device is connected to Arduino, the second is true when the user needs to keep the pet door closed regardless of the light and temperature values.

Setup code

We start setting up the Serial communication library:

```
Serial.begin(9600);
while (!Serial); // Leonardo/Micro should wait for serial init
```

This is not strictly necessary but it is required for writing to the console, which may be useful for debugging purposes.

Then we initialize the sensor library:

```
sensors.begin();
```

We also read the address of the device number 0 (the only one present in our circuit):

```
if (!sensors.getAddress(temperatureSensorAddress, 0))
    Serial.println("Unable to find address for Device 0");
```

The 1-Wire address

Each 1-Wire device has its own address defined at time of production and it cannot be changed. To perform any operation on the device you have to know its address. The getAddress and search library functions can help you to find the address of your devices.

The sensor can provide readings with different levels of precision, but the more precision we ask for, the more the device is slow to answer. For our purposes, we don't need a high precision, so we can set the precision to 9 bits:

```
sensors.setResolution(temperatureSensorAddress, 9);
```

The library has to know the pin where the servo motor is wired on:

```
myservo.attach(SERVOPIN);
```

Both the pins used for the photoresistor and the switch need to be configured as input:

```
pinMode(PHOTORESISTORPIN, INPUT);
pinMode(SWITCHPIN, INPUT);
```

For communicating with the nRF8001 board we need to set up a couple of callback functions; one, to know when the iOS device is connected or disconnected:

```
uart.setACIcallback(aciCallback);
```

The other callback function is for receiving data that are transmitted by the iOS device:

```
uart.setRXcallback(rxCallback);
```

The nRF8001 library calls the rxCallback when data from the iOS device are available for processing.

We are now ready to take a look at the main part of the code, which implements the algorithm to control the pet door and the iOS communication.

Main program

From the project's requirements, we conclude that the Arduino program has to implement a simple logic:

- Read light intensity
- Read temperature
- If the light intensity is above `LIGHT_THRESHOLD`, at a temperature between `LOW_TEMPERATURE_THRESHOLD` and `HIGH_TEMPERATURE_THRESHOLD`, the servo motor has to move to the `UNLOCK_POSITION` (180 degree), otherwise to the `LOCK_POSITION` (65 degree)

Moreover, when the iOS device is connected, it has to receive information about the latch position (opened or closed) and the external temperature.

In the main loop function we read the light intensity with:

```
unsigned int light = analogRead(PHOTORESISTORPIN);
```

The temperature with:

```
boolean lacthIsOpened = digitalRead(SWITCHPIN);
```

And the latch position with:

```
sensors.requestTemperatures();
float temperature = sensors.getTempC(temperatureSensorAddress);
```

Light is a value in the range 0-1023 (2^{10}) proportional to the light intensity, if the latch is opened, the `latchIsOpened` has value true (1), and `temperature` is the temperature measured by the sensor.

Voltage at the analog pin:

The **analog to digital converter** (**ADC**) inside Arduino, attached to an analog pin, transforms the voltage at the pin into an integer value using 10 bits, which is an integer value in the range 0-1023 (2^{10} values).

Since the voltage at the analog pin can be between 0 V and 5 V (the power voltage) each bit has a value that is 5 / 1024, then you can calculate the voltage at the analog pin using this formula:

*voltage = analogRead(<ANALOG PIN>) * 5/1024*

If you power on Arduino through the USB port of your computer, the power voltage is never exactly 5V but usually less. To get a better reading, you should measure the power voltage with a digital multimeter, and replace 5 with the actual power voltage in the previous formula.

In the loop function it's necessary to call the `pollACI` function so that the communication library can take the control for handling incoming data from the iOS connected device.

> **poolACI**
>
> If you have a long and complex program, you may need to add many `poolACI` calls across the code in order to frequently allow the library to take control to handle communication. Otherwise, you can experience data loss.

The core algorithm is in the following lines:

```
if (!manuallyLocked) {

    if (aboveThreshold(light, LIGHT_THRESHOLD, 30) &&
betweenThresholds(temperature, LOW_TEMPERATURE_THRESHOLD, HIGH_
TEMPERATURE_THRESHOLD)) {
        Serial.println("Unlocked");
        myservo.write(UNLOCKED_POSITION);
    }

    if (belowThreshold(light, LIGHT_THRESHOLD, 30) ||
!betweenThresholds(temperature, LOW_TEMPERATURE_THRESHOLD, HIGH_
TEMPERATURE_THRESHOLD)) {
        Serial.println("Locked");
        myservo.write(LOCKED_POSITION);
    }

}
```

If the door has not been manually locked, the light is above the `LIGHT_THRESHOLD`, the temperature is between `LOW_TEMPERATURE_THRESHOLD` and `HIGH_TEMPERATURE_THRESHOLD`, the servo motor can be moved to the `UNLOCKED_POSITION` with the instruction:

```
myservo.write(UNLOCKED_POSITION);
```

Otherwise, it is moved to the `LOCKED_POSITION`:

```
myservo.write(LOCKED_POSITION);
```

DeMorgan's theorem

This theorem is very useful for coding if-then-else statements in any programming language. Don't ever forget it!

not (A or B) = not A and not B

or

not (A and B) = not A or not B

The opposite condition of "A or B" is "not A **and** not B", the opposite condition of "A and B" is "not A **or** not B.

The following two functions are quite self-explanatory:

- aboveThreshold(...)
- betweenThresholds(....)

Take a look at the downloaded code for more details.

If manuallyLocked is true, light and temperature readings are ignored. This variable is set by a message received from the iOS device in the rxCallback function:

```
void rxCallback(uint8_t *buffer, uint8_t len) {

  if (len > 0) {

    // Data received from the iOS device
    // Received only one byte which has value 48 (character 0) or 49
(character 1)

    manuallyLocked = buffer[0] - '0';
    if (manuallyLocked) {
      Serial.println("Manual Lock");
      myservo.write(LOCKED_POSITION);
    }

  }
}
```

Remember that the nRF8001 library automatically calls the rxCallback function when data sent from the iOS device are ready for processing.

The iOS device sends only one byte with the ASCII character 1 if the door has to be locked, or the ASCII character 0 otherwise.

The ASCII character 0 has code 48, so to transform it to a Boolean value false (0) you need to subtract 48 (or the character '0' which is the same). Subtracting 48 from the ASCII character 1, we get the Boolean value true (1).

Back to the main function. If iOSConnected is true, an iOS device is connected to Arduino, and some data has to be transferred to it.

Data are sent to iOS with the following format:

```
s:latch_position;t:temperature
```

Here, latch_position informs the iOS device if the latch is opened or closed, whereas temperature is the external temperature.

To send data to iOS we use the following code:

```
if (iOSConnected) {

    // When the iOS device is connected some data are transferred to
it

    char buffer[32];
    char tempBuffer[6];

    // Data sent to iOS
    // s:latch_position;t:temperature

    dtostrf(temperature, 0, 2, tempBuffer);

    snprintf(buffer, 32, "s:%d;t:%s", lacthIsOpened, tempBuffer);
    uart.write((uint8_t *)buffer, strlen(buffer));
}
```

The function snprintf creates a buffer formatted as required, which is then sent to iOS with the function uart.write((uint8_t *)buffer, strlen(buffer)).

snprintf

This function never writes more characters than that indicated by the second parameter. This is very important to write safe code. In fact, if you write more characters than the buffer size, it is likely that you write somewhere in a memory location which is used for other purposes causing a microprocessor crash. After a crash, the microprocessor restarts the program execution from the setup function. For more details, visit http://bit.ly/1EO21no and http://bit.ly/1LijDx5.

dtostrf

Unfortunately, on Arduino, snprintf doesn't work with floating numbers, so we need to use dtostrf to transform the temperature (which is a float) to a string and then using it in the snprintf. The second parameter of dtostrf is the number of decimal digits to use in the conversion. For more details, visit: http://bit.ly/1fmj9HV.

The last thing that we have to code is managing connections and disconnections of the iOS device. This is done in the aciCallback function. The nRF8001 library calls this function each time one of the following events happen: the device starts advertising itself in order to be discovered by other Bluetooth devices, an external device connects, or a connected device disconnects.

```
void aciCallback(aci_evt_opcode_t event) {

  if (event == ACI_EVT_DEVICE_STARTED)
    Serial.println(F("Advertising started"));

  if (event == ACI_EVT_CONNECTED) {

    iOSConnected = true;

    char buffer[16];
    snprintf(buffer, 16, "m:%d", manuallyLocked);
    uart.write((uint8_t *)buffer, strlen(buffer));
  }

  if (event == ACI_EVT_DISCONNECTED) {
    iOSConnected = false;
  }
}
```

Testing and tuning the Arduino side

Once you have uploaded the code to Arduino (for more details: `http://bit.ly/1JPNAn3` or `http://bit.ly/1KsUhqv`), you can start testing it.

If enough light reaches the photoresistor and the temperature is between 2 degree Celsius and 33 degree Celsius, the servo should move to the open position and the latch should be completely retracted.

Then, when you cover the photoresistor, the latch should close. In order to test the temperature sensor you can use a hairdryer to get the temperature over 3 degree Celsius and an ice pack to get it below 2 degree Celsius. In both cases the latch should close.

You can easily change the temperature thresholds by changing these defines:

```
#define LOW_TEMPERATURE_THRESHOLD    2
#define HIGH_TEMPERATURE_THRESHOLD   33
```

The light threshold likely requires some more tuning, since the values read by the photoresistor highly depend on the photoresistor characteristics, the mounting positioning and orientation. Anyway, change:

```
#define LIGHT_THRESHOLD            700
```

You should be able to find the value that is right for your needs.

You may also need to adjust the servo motor positions for opening and closing the latch because of the way you have assembled everything. To adjust the positions you can change:

```
#define LOCKED_POSITION          180
#define UNLOCKED_POSITION         65
```

This represents the position in degrees of the motor.

iOS code

In this chapter, we are going to write the iOS application through which we can connect to the Arduino, know if the pet door is locked or not, read the external temperature, and eventually manually lock it.

The full code of this project can be downloaded from here:

```
https://www.packtpub.com/books/content/support
```

To better understand the explanations in the next paragraphs, you should open the downloaded code while reading.

The main tool for writing the application is Xcode provided by Apple. We can start it from Launchpad or entering the **Applications** folder in **Finder**.

Start developing iOS apps today

A useful guide to iOS developing provided by Apple can be found at this link: http://apple.co/MtP2Aq.

Objective-C

The language used for developing iOS applications is Objective-C. It is similar to C++ and you can find an introduction to this language at the following link: http://apple.co/19FWxfQ.

Creating the Xcode project

The first step is to create a new project. Xcode provides many different project templates; we are going to use the Tabbed Application, which has two tabs. We are going to use the first tab as the main application panel, and the second to scan the nRF8001 device. This operation is required only once, when the application is started for the first time.

Xcode overview

You can find everything you need to work with Xcode at the following link: http://apple.co/1UQnMtS.

To create the new project we can follow these steps (see the following screenshots):

1. Go to **File | New | Project ....**
2. In the left panel, select **iOS | Application**.
3. In the right panel, select **Tabbed Application** and then click on **Next**.

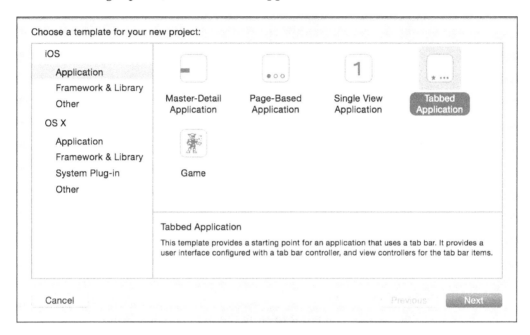

4. In the next screen, enter the required information:
    ◦ **Product Name**: PetDoorLocker
    ◦ **Organization Name**: Your Name
    ◦ **Organization Identifier**: yourname (this information is relevant only for publishing the application to the iTunes Store and selling it. It can be ignored)
    ◦ **Language: Objective-C**

- ° **Devices**: **Universal** (we are going to create an application that can run on an iPhone and iPad)

5. Click on the **Next** button.
6. Select a folder where you want to store the project (leave Source Control unchecked).

That's all!

We are now ready to start writing the new application. The first thing to do is rename the two view controllers (FirstViewController and SecondViewController).

1. Select `FirstViewController.h` in the left panel, this will open the file in the right panel.
2. Select `FirstViewController` in the line `@interface FirstViewController : UIViewController`, by double clicking on `FirstViewController`.
3. Right Click and select **Refactor | Rename ....**
4. Enter the new name of the view controller: `PetDoorLockerViewController`.
5. Click preview and then save.
6. Select `SecondViewController.h`, and with the same procedure rename it as `BLEConnectionViewController`.

What's a View Controller? The Apple documentation says:

> *View controllers are a vital link between an app's data and its visual appearance.*
> *Whenever an iOS app displays a user interface, the displayed content is managed*
> *by a view controller or a group of view controllers coordinating with each other.*
> *Therefore, view controllers provide the skeletal framework on which you build*
> *your apps.*

Design Patterns
Before going any further, we suggest you read the information
on the following link: http://apple.co/1hkUDbU.

Now we are ready to design the GUI of the new application.

Designing the application user interface for BLEConnectionViewController

To design the app user interface, let's open the Main.storyboard. This file contains everything about the GUI. Once opened, you should see something as shown in the following screenshot:

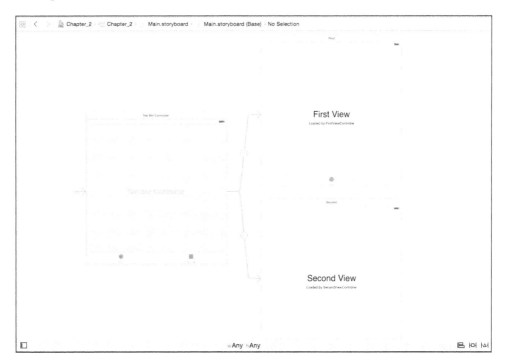

Let's start from the BLEConnectionViewController, which will be used for scanning the nRF8001 device.

1. Double click on the view controller to select it.

2. Click on the label **Second View** and delete it, then select **Loaded by the SecondViewController** and delete it.

3. Open the Utilities panel on the right: **View | Utilities | Show Utilities**. (To open this panel you can also use the green circled icon in the following screenshot).

4. Select **Label** in the Tool Navigator and drop it to the free area (see the following screenshot).

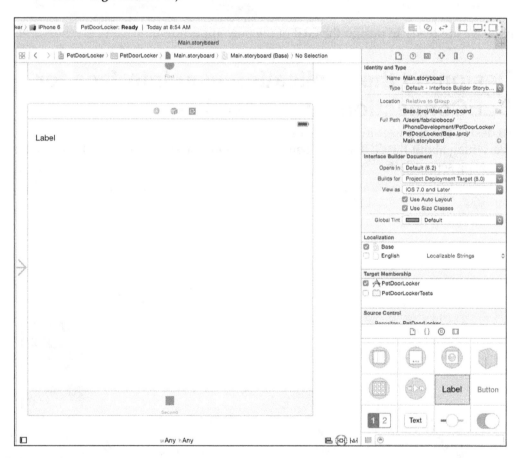

5. Rename the label, by double clicking **Label** and entering `Device`.

6. Now we have to set the Auto Layout constraints that will lock the label in the desired position:

    1. Click on the Auto Layout Pin icon (it's circled in red in the previous screenshot).

    2. Enter `20` for the Leading Space from Superview, and `30` for Top Space to Superview as shown in the following screenshot.

    3. Enter `53` as **Width**.

    4. Click on **Add 3 Constraints**.

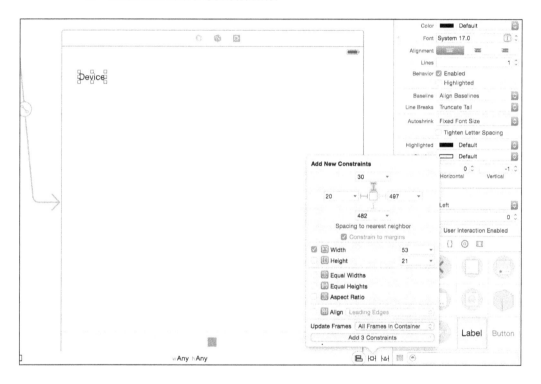

7. Drop a new instance of **Label** next to the previous and add Auto Layout Constraints to it:

    1. Select both labels (clicking on them with the *Command* button pressed).

    2. Click on the Align icon (it's circled in green in the previous screenshot).

3. Select **Vertical Centers** and then **Add 1 Constraint**.

4. Select only the new label and click on the Auto Layout Pin.

5. Enter 20 for both the Leading Space and Trailing Space.

6. For **Update Frames**, select **All frames in Container**.

7. Click on **Add 2 Constraints**.

8. Select the new label again, and select **View | Utilities | Show Attributes Inspector** (or click on the icon circled in red in the next screenshot).

9. Change the font size to 13 and **Alignment** to center (see area circled green in the next screenshot).

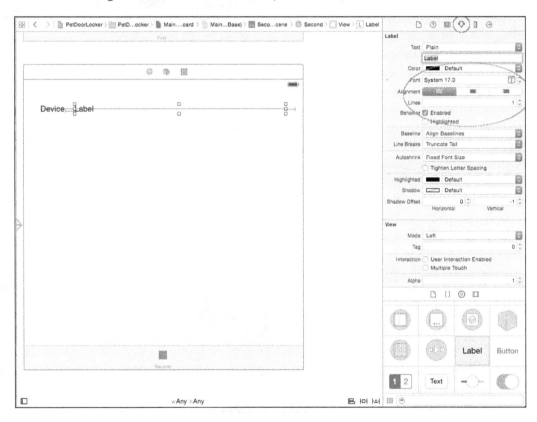

The new label will show the UUID of the nRF8001 once detected. Now we have to add a button to start scanning for nRF8001 devices nearby.

8. Drag a button into the container, double click on it, and enter `Scan`:
    1. Select the button and click on the Auto Layout Pin.
    2. Enter `45` for **To Space** and click on **Add 1 Constraint**.
    3. Select the button, click on the Align icon, select Horizontal Center in the Container, for **Update Frames**, select **All frames in Container**, and then click on **Add 1 Constraint**.

    Now we have to link the GUI components with the code in order to manipulate them programmatically.

9. Select the **BLEConnectionViewController**.

10. Click on **View | Assistant Editor**. A new panel opens with the `BLEConnectionViewController.h`.

11. Close the Utilities panel to have more space (click on **View | Utilities | Hide Utilities**).

12. Make sure that in the red circled area of the following screenshot, you read **BLEConnectionViewController.m**, otherwise click on it and change to the desired file.

13. Select the label **Label** and keeping the *Command* button pressed, drag the label to the code on the right between `@interface BLEConnectionViewController ()` and `@end` (see picture below).

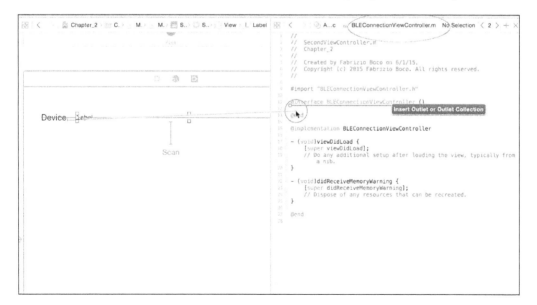

14. When a dialog appears, enter the **Name** as `deviceUUIDLabel` (see picture below) and then click **Connect**.

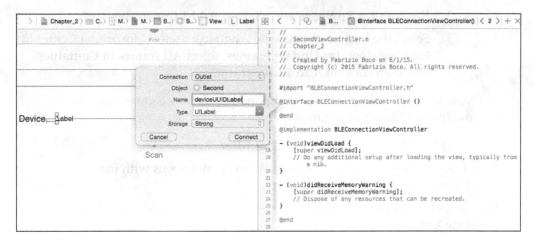

This creates a property (`deviceUUIDLabel`), which can be used for changing the label properties, such as its text.

ARC and strong versus weak

The `strong` attribute indicates to the compiler that the memory associated to the property has to be kept allocated until the class in which it is defined is allocated. Properties that are not defined strong (but weak) are automatically deallocated as soon as the code block in which they are defined is completed. In short, as long as there is a strong reference pointer to the object, that object will not be deallocated. Memory management under iOS has been simplified since the introduction of **Automatic Counting Reference** (**ARC**) but is still an issue for most people. A good introduction can be found here: `http://apple.co/1MvuNgw`. It is for Swift (the last programming language that Apple has made available) but it is worth reading.

Atomic versus nonatomic

The `atomic` attribute will ensure that a whole value is always returned from the getter or set by the setter of the property, regardless of setter activity on any other thread. That is, if thread A is in the middle of the getter while thread B calls the setter, an actual viable value will be returned to the caller in A.

The Apple documentation says: *Property atomicity is not synonymous with an object's thread safety.*

Accessing a strong property is slower than accessing a nonatomic one. More details here: `http://apple.co/1JeBIdb`.

We now link the **Scan** button to have a method that is called when the button is tapped:

15. Select the **Scan** button and, keeping *Control* pressed, drop it to the right pane.

16. Enter `startScanning` as the name and select **UIButton** for **Type** (see picture below). This creates a new method, which is called when the button is pressed.

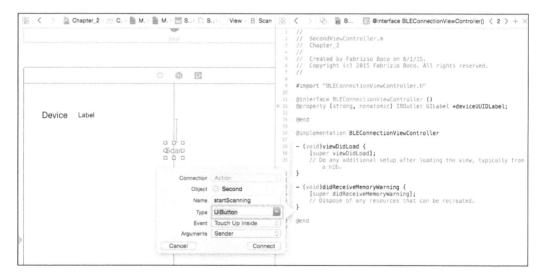

To complete the design, we have to set the name of the view controller, which appears in the toolbar:

17. Open the Tool Navigator again (**View | Utilities | Show Utilities**).

18. Select the icon at the bottom side of the container (a small square with second label under it).

19. Show Attribute Inspector (**View | Utilities | Attribute Inspector**).

20. In the **Title** field, enter `Configuration`.

> You could also choose an icon from the Image list box. To do that, you should add the icon to the project, dropping it into the Supporting Files group in the left panel. The icon should be 32 × 32 pixels.

At the very end, your View Controller should look like the next picture.

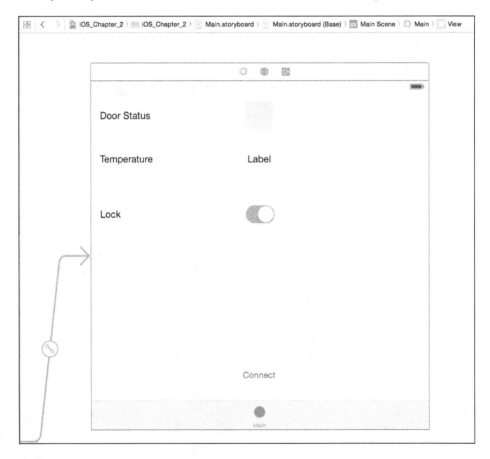

Designing the application user interface for PetDoorLockerViewController

In this section we will design the interface for the first view controller (PetDoorLockerViewController). We will only describe how to add components not shown in the previous chapter. Refer to the next picture for the global layout of the view.

1. Add a label, rename it as `Door Status` and add to it the Auto Layout constraints.

2. Enter `UIView` in the search field of the right panel and drop the view near the label. You can resize the view dragging the small white squares on its borders. In the Attribute Inspector, select a light gray color for the background.

3. Select both the label and the view and click on Align icon. Select **Vertical Centers** and click on **Add 1 Constraint**.

4. Select the view and click on the Align icon. Select **Horizontal Center** in the **Container** and click on **Add 1 Constraint**.

5. Select the Pin icon, enter `48` for height and width, for **Update Frames**, select **All frames in Container** and then click on **Add 2 Constraints**.

6. Add a new label `Temperature`, and add to it the Auto Layout constraints.

7. Add another label near **Temperature** and the Auto Layout constraints to vertically center it with **Temperature** and horizontally center it in the container's view.

8. Then add a label, `Lock`, and add Auto Layout constraints.

9. Add a switch near **Lock** and the Auto Layout constraints to vertically center it with **Lock** and horizontally center it in the container's view.

10. At the bottom of the container, add a button (called **Connect**) and the Auto Layout constraints to vertically center it with **Lock** and horizontally center it in the container's view.

You should end with something like what is shown in the following picture.
The **Connect** button is useful in case you lose connection to the nRF8001 device,
and you need to manually reconnect to it.

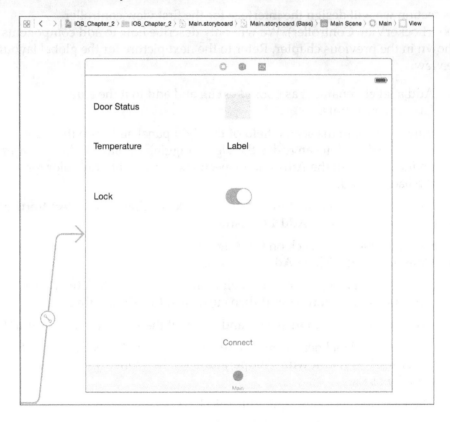

Once all the components have been added, you can link them to the code
finishing with:

```
@interface PetDoorLockerViewController ()

@property (strong, nonatomic) IBOutlet UIView      *doorStatus;
@property (strong, nonatomic) IBOutlet UILabel     *temperature;
@property (strong, nonatomic) IBOutlet UISwitch    *manualLockSwitch;


@end
```

And with the function `reconnect` to intercept when the **Connect** button is tapped.

In case of doubts, you can use the downloaded code as reference.

We are eventually ready to write the code for both the view controllers.

Writing code for BLEConnectionViewController

The purpose of this controller is to get the unique identifier of the nRF8001 device, so that it can be used to connect to the device as needed.

To handle communication with a Bluetooth 4.0 device, we need to use the class `CBCentralManager` that we add to the interface of the view controller:

```
@interface BLEConnectionViewController ()

@property (strong, nonatomic) IBOutlet UILabel  *deviceUUIDLabel;

@property (strong, nonatomic) CBCentralManager  *centralManager;


@end
```

Xcode class reference

If you need to know more about a class, you can press *Option* + click the class name for direct access to the documentation. Conversely, *Command* + click will bring you to the source file.

The instantiation of the class is in the `viewDidAppear` method, which is called every time the view associated with the view controller is shown on the device screen.

```
-(void)viewDidAppear:(BOOL)animated {

    _centralManager = [[CBCentralManager alloc] initWithDelegate:self
queue:nil];
}
```

To work with `CBCentralManager`, we need to implement a few delegate methods, but first we have to instruct the controller about that; we open the `BLEConnectionViewController.h` file and we change it as follows:

```
#import <UIKit/UIKit.h>
#import <CoreBluetooth/CoreBluetooth.h>

@interface BLEConnectionViewController : UIViewController
<CBCentralManagerDelegate>


@end
```

Then we are ready to write two delegate methods:

- `centralManagerDidUpdateState`: This method is called when the state of the iOS Bluetooth subsystem changes. This method has only an informative purpose in this view controller, but it is very useful in more complex projects where the status of the Bluetooth subsystem needs to be monitored. See the downloaded code to get more details about this method.

- `didDiscoverPeripheral`: This method is called when a new peripheral is discovered:

```
- (void)centralManager:(CBCentralManager *)central didDiscoverPeri
pheral:(CBPeripheral *)peripheral advertisementData:(NSDictionary
*)advertisementData RSSI:(NSNumber *)RSSI {

    [_scanningTimer invalidate];
    _deviceUUIDLabel.text = peripheral.identifier.UUIDString;

    NSUserDefaults *userDefaults = [NSUserDefaults
standardUserDefaults];

    [userDefaults setObject:peripheral.identifier.UUIDString
                forKey:@"PetDoorLockerDevice"];

    [userDefaults synchronize];
}
```

Each time a new Bluetooth peripheral is discovered (the nRF8001 board hooked up to Arduino in our case) the method is called providing information on the peripheral. We will show the identifier of the peripheral on the screen:

```
_deviceUUIDLabel.text = peripheral.identifier.UUIDString;
```

And we store it in `userDefaults` with the key `PetDoorLockerDevice` so that we can retrieve it as needed.

User preferences

Preferences are information that you store persistently, and use to configure your app. They can be accessed using the `NSUserDefaults` class. More details can be found here: `http://apple.co/1qRYb3o`.

Later on, we'll get back to explaining:

```
[_scanningTimer invalidate];
```

The **Scan** button activates the method `startScanning` when the user needs to detect the available device.

```
- (IBAction)startScanning:(UIButton *)sender {

    if (_centralManager.state != CBCentralManagerStatePoweredOn)
        return;

    [_centralManager scanForPeripheralsWithServices:@[[CBUUID
UUIDWithString:NRF8001BB_SERVICE_UUID]] options:nil];

    _deviceUUIDLabel.text = @"Scanning...";

    _scanningTimer = [NSTimer scheduledTimerWithTimeInterval:(flo
at)5.0 target:self selector:@selector(scanningTimedOut:) userInfo:nil
repeats:NO];
}
```

If the state of the `centralManager` is not `CBCentralManagerStatePoweredOn` nothing can be done. Otherwise the scanning is started by calling `scanForPeripheralsWithServices`.

Each Bluetooth 4.0 device has one or more services uniquely identified; we look for the service identifier of the nRF8001 board (NRF8001BB_SERVICE_UUID).

nRF8001 service and characteristics

A Bluetooth peripheral can provide more services and, for each service, more characteristics that the user can read and/or write. The nRF8001 board has only one service (UUID: 6E400001-B5A3-F393-E0A9-E50E24DCCA9E), one characteristic to receive data (UUID: 6E400002-B5A3-F393-E0A9-E50E24DCCA9E) and one for sending data (UUID: 6E400003-B5A3-F393-E0A9-E50E24DCCA9E).

The value NRF8001BB_SERVICE_UUID is defined at the beginning of the view controller's code: `#define NRF8001BB_SERVICE_UUID @"6E400001-B5A3-F393-E0A9-E50E24DCCA9E"`.

Once a device with that service has been found, the iOS calls `didDiscoverPeripheral` and the iOS device stops scanning. Unfortunately, the scanning process runs until a device is found. So in case the peripheral has not been found, the iOS device keeps draining the batteries. To overcome this issue, we need a timer. It is defined into the view controller interface:

```
@interface BLEConnectionViewController ()

@property (strong, nonatomic) IBOutlet UILabel       *deviceUUIDLabel;

@property (strong, nonatomic) CBCentralManager       *centralManager;
@property (strong, nonatomic) NSTimer                *scanningTimer;

@end
```

And instantiated with:

```
_scanningTimer = [NSTimer scheduledTimerWithTimeInterval:(float)5.0
target:self selector:@selector(scanningTimedOut:) userInfo:nil
repeats:NO];
```

If not halted, the timer will call the method `scanningTimedOut` after 5 seconds. In this method we can stop `centralManager` from scanning and draining the batteries:

```
-(void) scanningTimedOut:(NSTimer *)timer {

    [_centralManager stopScan];
    _deviceUUIDLabel.text = @"No device in range";
}
```

If a peripheral, with the desired service, is found, the iOS calls the `didDiscoverPeripheral` method:

```
- (void)centralManager:(CBCentralManager *)central didDiscoverPerip
heral:(CBPeripheral *)peripheral advertisementData:(NSDictionary *)
advertisementData RSSI:(NSNumber *)RSSI {

    [_scanningTimer invalidate];
    _deviceUUIDLabel.text = peripheral.identifier.UUIDString;
```

```
    NSUserDefaults *userDefaults = [NSUserDefaults
standardUserDefaults];

    [userDefaults setObject:peripheral.identifier.UUIDString
                  forKey:@"PetDoorLockerDevice"];

    [userDefaults synchronize];
}
```

We have to halt the `scanningTimer` that is the purpose of this line `[_scanningTimer invalidate];` and save the UUID of the peripheral in User Defaults.

Now it's time to run the app on your device.

Run your app

All the information for running your app both on the simulator and on the physical device can be found here: `https://developer.apple.com/library/ios/documentation/IDEs/Conceptual/AppDistributionGuide/LaunchingYourApponDevices/LaunchingYourApponDevices.html`.

Power on the Arduino board and start the app, then tap on the configuration tab and then on the **Scan** button. In a few seconds you should see a long string of letters and numbers near the device label. That is the UUID of the nRF8001 device. The setup configuration has been completed!

If something goes wrong, the message **no device is in range** appears. In that case, double check the following:

1. The nRF8001 is properly wired to Arduino.
2. You have uploaded the correct code to Arduino.
3. The Arduino board is powered up.
4. The IDE console shows the messages **Setup Completed** and **Advertising started**.
5. Bluetooth is activated on your iOS device (tap Settings and then Bluetooth to activate it).

Writing code for PetDoorLockerViewController

In this section, we will write the main part of the application, which allows you to monitor if the pet door is locked or unlocked, read the external temperature, and lock the pet door if you need it.

 Since the code of this view controller is more complex and we want to save space for other subjects, we are going to explain the entire code but we are not going to guide you on writing it step-by-step. Please, refer to the downloaded code to see the code in its entirety.

We need three defines and some additional properties in the view controller's interface:

```
#define NRF8001BB_SERVICE_UUID              @"6E400001-B5A3-
F393-E0A9-E50E24DCCA9E"
#define NRF8001BB_CHAR_TX_UUID              @"6E400002-B5A3-
F393-E0A9-E50E24DCCA9E"
#define NRF8001BB_CHAR_RX_UUID              @"6E400003-B5A3-
F393-E0A9-E50E24DCCA9E"

@interface PetDoorLockerViewController ()

...

@property (strong, nonatomic) CBCentralManager      *centralManager;
@property (strong, nonatomic) CBPeripheral          *arduinoDevice;
@property (strong, nonatomic) CBCharacteristic
*sendCharacteristic;

@end
```

We also need to add a delegate `@interface PetDoorLockerViewController : UIViewController <CBCentralManagerDelegate, CBPeripheralDelegate>` in the `PetDoorLockerViewController.h`.

Setting up the `CBCentralManager` instance is exactly like we did in the previous controller.

The method `centralManagerDidUpdateState` is quite different:

```
- (void) centralManagerDidUpdateState: (CBCentralManager *) central {

    NSLog(@"Status of CoreBluetooth central manager changed %ld (%s)",
central.state, [self centralManagerStateToString:central.state]);

    if (central.state == CBCentralManagerStatePoweredOn) {

        [self connect];
    }
}
```

As soon as the Bluetooth subsystem is ready (its state is
`CBCentralManagerStatePoweredOn`) the app starts trying to connect to the nRF8001
board, calling `[self connect]` which is a method that we'll show you very soon.

The connection is also started from `viewDidAppear` each time the view controller is
shown on the screen:

```
- (void) viewDidAppear: (BOOL) animated {

    [super viewDidAppear:animated];

    [self connect];
}
```

The connection is closed as soon as the view disappears from the screen to reduce
draining of the batteries:

```
- (void) viewDidDisappear: (BOOL) animated {

    [super viewDidDisappear:animated];

    [self disconnect];
}
```

Now let's take a closer look at the connection method.

```
- (void) connect {

    if (_arduinoDevice == nil) {
```

```
        // We need to retrieve the Arduino peripheral

        NSString *deviceIdentifier = [[NSUserDefaults
standardUserDefaults] objectForKey:@"PetDoorLockerDevice"];

        if (deviceIdentifier!=nil) {

            NSArray *devices = [_centralManager
retrievePeripheralsWithIdentifiers:@[[CBUUID UUIDWithString:deviceIde
ntifier]]];
            _arduinoDevice = devices[0];
            _arduinoDevice.delegate = self;
        }
        else {


            ...
            ...

            return;
        }
    }

    [_centralManager connectPeripheral:_arduinoDevice options:nil];
}
```

If `arduinoDevice` is not initialized, we retrieve it using the UUID that has been
stored in the user preferences during the scanning phase. It's important that the
peripheral delegate is set, because we have to discover the peripheral's characteristics
and they are returned through delegate methods. The method `connectPeripheral`
actually connects to the peripheral. If the connection is successful, the delegate
method `didConnectPeripheral` is called, and we can start discovering the service
that the device provides:

```
- (void)centralManager:(CBCentralManager *)central didConnectPeriphera
l:(CBPeripheral *)peripheral {

    [peripheral discoverServices:@[[CBUUID UUIDWithString:NRF8001BB_
SERVICE_UUID]]];

}
```

Discovering all services

In certain cases you may need to discover all the services that a peripheral provides. To do that, you use: `[peripheral discoverServices:nil];`.

Once iOS discovers the peripheral's service, it calls the method `didDiscoverServices` and we can start discovering the characteristics of the service:

```
- (void)peripheral:(CBPeripheral *)peripheral
didDiscoverServices:(NSError *)error {

    ...

    for (int i=0; i < peripheral.services.count; i++) {

        CBService *s = [peripheral.services objectAtIndex:i];
        [peripheral discoverCharacteristics:nil forService:s];
    }
}
```

For each service provided, the iOS calls the method `didDiscoverCharacteristicsForService` (please see downloaded code). In this method we store the characteristic for sending data to the nRF8001 device in the property `sendCharacteristic`, and we call this method:

```
[peripheral setNotifyValue:YES forCharacteristic:characteristic];
```

Now, we'll use the characteristic for receiving data as parameter. Now each time the characteristic changes (data are sent from the nRF80001 device), the method `didUpdateValueForCharacteristic` is called and the available data are received (please, see downloaded code).

When incoming data are available, the `dataReceived` method is called and received data can be processed:

```
- (void)dataReceived:(NSString *)content {

    // Messages has the following formats:
    //
    //   1) m:0|1
    //
```

```
        // 2) s:0|1;t:temperature

    NSArray *messages = [content componentsSeparatedByString:@";"];

    for (int i=0; i<messages.count; i++) {

        NSArray *components = [messages[i]
componentsSeparatedByString:@":"];

        NSString *command = components[0];
        NSString *value   = components[1];

        if ([command isEqualToString:@"m"]) {
            _manualLockSwitch.on = [value boolValue];
        }

        if ([command isEqualToString:@"s"]) {

            BOOL doorUnlocked = [value boolValue];

            if (doorUnlocked) {

                _doorStatus.backgroundColor = [UIColor greenColor];
            }
            else {

                _doorStatus.backgroundColor = [UIColor redColor];
            }
        }

        if ([command isEqualToString:@"t"]) {
            _temperature.text = value;
        }
    }
}
```

We can receive two kinds of messages:

- m:0|1
- s:0|1; t:temperature

When the iOS device connects to Arduino, it receives the first message, which informs you if the door has been manually locked (m:1) or not (m:0). With this information we can set the position of the manual switch:

```
_manualLockSwitch.on = [value boolValue];
```

The second message contains two types of information: if the latch is open (s:1) or closed (s:0) and the external temperature. The first is used to change the background color of the doorStatus view:

```
BOOL doorUnlocked = [value boolValue];

if (doorUnlocked) {

    _doorStatus.backgroundColor = [UIColor greenColor];
}
else {

    _doorStatus.backgroundColor = [UIColor redColor];
}
```

The temperature information is used to set the value of the temperature label:

```
if ([command isEqualToString:@"t"]) {
    _temperature.text = value;
}
```

When the switch manualLockSwitch is tapped, the method switchChanged is called and there we can transmit data to Arduino:

```
- (IBAction)switchChanged:(UISwitch *)sender {

    NSData* data;

    if (sender.on)
        data=[@"1" dataUsingEncoding:NSUTF8StringEncoding];
    else
        data=[@"0" dataUsingEncoding:NSUTF8StringEncoding];

    [_arduinoDevice writeValue:data forCharacteristic:_
sendCharacteristic type:CBCharacteristicWriteWithoutResponse];

}
```

To send data to a Bluetooth device, we write to the appropriate characteristic using the method `writeValue`. Since it accepts `NSData` values, we have to convert the string `"0"` or `"1"` to `NSData` using the method `dataUsingEncoding`.

We have almost completed the application. Once the iOS application is connected to Arduino, we need that it disconnects when it is sent to the background (for saving batteries). When it is brought to the foreground again, it automatically reconnects to Arduino.

To do that we make the `connect` method public and we write a new `disconnect` public method. To make the methods public we add a couple of lines to `PetDoorLockerViewController.h`:

```
- (void) connect;
- (void) disconnect;
```

The `disconnect` method is very simple:

```
- (void) disconnect {

    if (_arduinoDevice != nil) {
        [_centralManager cancelPeripheralConnection:_arduinoDevice];
        _doorStatus.backgroundColor = [UIColor lightGrayColor];
    }
}
```

Setting the background color of the `doorStatus` view to light gray, we can visually know if the iOS is connected to Arduino or not.

The very last method we have to write is `reconnect`, which doesn't require any explanation:

```
- (IBAction) reconnect:(UIButton *) sender {

    [self disconnect];
    [self connect];
}
```

In the `AppDelegate.m` file there are two methods, which are respectively called when the app enters in the background or gets back to the foreground:

* `applicationDidEnterBackground`
* `applicationWillEnterForeground`

In these methods we need a reference to the `PetDoorLockerViewController`. We can get it through the main application window.

```
- (void)applicationDidEnterBackground:(UIApplication *)application {

    UITabBarController *tabController = (UITabBarController *)_window.
rootViewController;
    PetDoorLockerViewController *petDorLockerController =
tabController.viewControllers[0];

    [petDorLockerController disconnect];
}


- (void)applicationWillEnterForeground:(UIApplication *)application {


    UITabBarController *tabController = (UITabBarController *)_window.
rootViewController;
    PetDoorLockerViewController *petDorLockerController =
tabController.viewControllers[0];

    [petDorLockerController connect];
}
```

Testing the iOS app

Now the application is completed and we can run it again on our iOS device. As soon as it starts, it should connect to Arduino and the door status indicator should turn to red if the door is locked, or to green if it is unlocked and you should see the temperature measured from the Arduino sensor.

When you tap on the lock switch, the door should immediately close and ignore light and temperature.

How to go further

The application we have developed can be improved in many ways; these are some suggested improvements that you can try yourself:

- Checking the presence of your pet in the house by counting door openings and their directions. A couple of magnetic switches should do the job of detecting the opening direction of the pet door.

- Detect your own pet using an RFID tag attached to their collar to avoid other pets being able to get into your house.

- Setting light threshold and temperature thresholds directly from the iOS device using a slider (UISliderView).

- Replace the temperature numeric indication with a more appealing graphical indicator like a gauge or a thermometer.

- Notify you when your pet goes through the pet door.

- Show temperature in Celsius and Fahrenheit.

Different types of sensors

Before ending this chapter we would like to give an overview of the existing type of sensors, their communication protocols with Arduino, and the pros and cons of using them.

Sensors can be categorized into two major families (analog and digital) based on the kind of signal they provide. An analog sensor usually provides a voltage, which is proportional to the quantity it is measuring. This voltage has to be converted in a number using an ADC. Arduino provides six analog pins and each of them has its own ADC. The photoresistor that we are using in our project is a typical analog sensor. Conversely, a digital sensor directly provides a numeric representation of the measured quantity, which can be directly used. The temperature sensor, which we are using in this project, is an example of a digital sensor.

Analog sensors are easier to use and cheaper, but they are very sensitive to power voltage fluctuations and electrical noise in the circuit. For these reasons the readings change a lot and it is usually needed to implement digital filters in code to smooth out the readings.

Digital sensors instead provide very stable readings and they are usually more precise. Unfortunately, they communicate with the microprocessor using different low-level protocols, which are more complex to handle. In the majority of the cases, protocol complexities are hidden by software libraries specialized for each type of sensor, but this makes coding more complex and libraries usually lead to a greater memory consumption which is a very precious resource on a microprocessor.

The most used low-level protocols are **Serial Peripheral Interface (SPI)** and **Inter-integrated Circuit (I2C)**. Another low-level protocol largely used for temperature sensors is the 1-Wire, which has been adopted for this project.

A complete comparison between these protocols is out of the scope of this project, but you can get an idea of them by referring to the following table.

Protocol	Architecture	Signals needed	Multi-master	Data rate	Full duplex
SPI	Two shared uni-directional data signals and a shared clock	SCK, MISO, MOSI, and one CS for each device on the board	Possible, but not standard	1 Mbps	Yes
I2C	Shared data signal and a shared clock signal	SDA and SCL	Yes	100 kbps, 400 kbps, and 3.2 Mbps	No
1-Wire	One data signal	Data	No	15 kbps	No

Summary

Well done! You made it to the end of the chapter and you have built a project from scratch!

You have built the hardware, along with the electronic circuit, and written the software for both Arduino and iOS.

On Arduino, you have learned how to use analog and digital sensors (1-Wire), how to write the code for reading them, how to control a servo motor, and how to handle the communication with the iOS devices.

On iOS, you have learned how to write an application with a simple user interface and that the application's communication with Arduino via Bluetooth 4.0.

We eventually discussed analog and digital sensors and some of the most used low-level communication protocols for exchanging data with Arduino.

In the next chapter, we will build another project, which uses Wi-Fi instead of Bluetooth to transfer data. In that project, Arduino will accept different commands and react to it. The iOS application will have a table view, which is one of the most useful components provided by the UIKit.

3

Wi-Fi Power Plug

The Wi-Fi Power Plug is a device through which you can control electrical appliances hooked up to it, turning them on and off in two ways:

- Manually from your iOS device
- Automatically setting timers

For example, you can turn on your irrigation system every day at 6 P.M. for 30 minutes, but if you see that your grass is turning yellow you can manually turn on the system for an additional watering.

Nowadays, this kind of device is available on the market for a reasonable price but building one yourself will allow you to understand how it works and adapt it to your own needs.

In comparison with the project in *Chapter 1, Arduino and iOS – Platforms and Integration*, we are going to use Wi-Fi as the communication protocol. This will allow you to access the device even when you are not at home.

The chapter is organized in the following sections:

- Wi-Fi power plug requirements: We will briefly set the project requirements
- Hardware: We will describe the hardware and the electronic circuit needed for the project
- Arduino code: We will write the code for Arduino to control the external appliance and to communicate with the iOS device
- iOS code: We will write the code for the iOS device
- How to access your power plug from anywhere in the world
- How to go further: More ideas to improve the project and learn more

Wi-Fi power plug requirements

We are going to build a device that will be able to:

- Turn an electric appliance hooked up to it on and off by receiving a command from the iOS device

- Turn on and off an electrical appliance on and off at specific times for a predefined lasting

The accompanying iOS application has to manually control the power plug and manage the timers.

Hardware

As we mentioned in *Chapter 1, Arduino and iOS – Platforms and Integration*, we need a Wi-Fi shield (`http://www.arduino.cc/en/Main/ArduinoWiFiShield`) and a micro SD card formatted with FAT16 (take a look here for more details: `http://www.arduino.cc/en/Reference/SDCardNotes`). The SD card is used to permanently store the activation times when Arduino is powered off, so its size is not so important.

Additional electronics components

In this project we need some additional components:

- Optoisolator MOC3041

- 330 Ω resistor, 0.5 W

- 330 Ω resistor, 0.25 W

- Red LED

- TRIAC BTA08-600

The TRIAC is capable of 8 Amperes RMS at 600 V, which is about 1700 watts at 220 V. You can use another TRIAC model of the same family (for example, BTA16) if you have a more powerful external appliance.

Electronic circuit

The following picture shows the electric diagram of the electronic circuit that we need for the project:

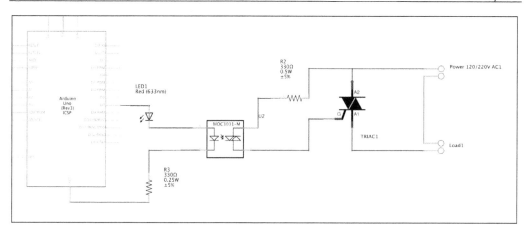

The following picture shows how to mount the circuit on a breadboard.

 Don't forget to mount the Wi-Fi shield, and insert a micro
SD card into it.

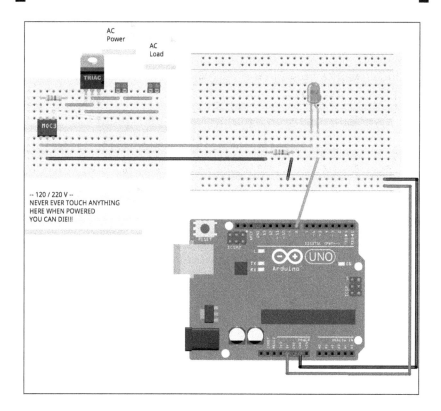

This circuit works with the power line (120 V or 220 V). Touching any part with power line voltage can be extremely dangerous and can kill you. Even professionals with years of experience have been injured or killed, so *please* be extremely careful. This means that if you have never worked with power line voltage, you need to do that under the control of a skilled person.

Again, avoid working with the power line yourself if you have not done it before. You are the only one responsible for any damage you may cause to yourself, your relatives and your stuff. You have been warned!

If you don't feel safe handling the power line, you can still enjoy the project by replacing the power circuit with just the LED and the resistor.

To power the external appliance we use a TRIAC (`http://bit.ly/1MzmIYs`), which allows you to control an AC load using a small current. Since the power plug uses the power line, the voltages around are high (120-220 V) and they can burn your Arduino. For that reason, an optoisolator is placed between the low voltage circuit (Arduino) and the high voltage circuit (TRIAC) (`http://bit.ly/1TV1JFc`). Basically, it is an LED and a low power TRIAC in the same small package. When the LED is on, a small current is generated for photoelectric effect and it polarizes the gate of the low power TRIAC switching it. The main point here is that the light electrically isolates the LED and the TRIAC.

A transistor and a relay can replace the TRIAC and the optoisolator, but a relay is an electromechanical device, which is subject to failures in short time.

Turning the optoisolator LED from Arduino on and off is pretty easy, as it is wired to a digital pin (number 8 in our case) and controlled with `digitalWrite(<PIN>, HIGH | LOW)`. The external LED in series with the optoisolator has monitoring purpose only.

Controlling more appliances

If you need to control more appliances, you can replicate the power circuit (Optoisolator and TRIAC) and wire it to another digital pin. Then you need to adapt both the Arduino code and the iOS code.

Arduino code

The full code of this project can be downloaded from here:

`https://www.packtpub.com/books/content/support`

For a better understanding of the explanations in the next paragraphs, you should open the downloaded code while reading.

The power plug has to activate the external appliance at different times for a different lasting and then deactivate it. We are going to call **activation** a turn on-turn off cycle. Each activation can be represented by a square wave (see the following diagram).

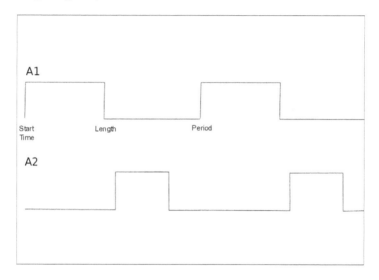

Each activation starts at its **Start Time**, lasts for a **Length** (during which the appliance is on) and repeats exactly after a **Period** of time. A one-off activation has a period equal to 0.

We will use this diagram for understanding the Arduino code.

The application can manage the NUMBER_OF_ACTIVATIONS activations stored in the global array `activations`. Each activation is a new type, defined as follows:

```
typedef struct {
  char                name[21];
  unsigned long       startTime;      // seconds since midnight
1/1/1970
  uint16_t      length;        // minutes
  uint16_t      period;        // minutes
} activation;
```

To make the code as simple as possible (and to save the flash memory) we save the entire array to the SD file, even if not all the activations are set. An activation is set if its name is set.

Flash memory

Flash memory stores the program running on Arduino. Arduino UNO has 32 K of flash memory of which 0.5 K is used for the bootloader. The bootloader is a small code that allows programming UNO via USB instead of using an external in-circuit programmer (for example, an AVR-ISP or STK500).

Setup code

Please refer to the downloaded code, since the setup code is quite simple and doesn't require a detailed explanation.

Because we are handling time and Arduino does not have a real-time clock, we need to get the current time from a Network Time Protocol server on the net (`http://bit.ly/1NyCVM7`); this is what the function `askTime` does. The request is sent via UPD (`http://bit.ly/1MzmQHb`) and the answer is received on port 2390. When a data packet is received, it is transformed into Unix time with the function `readTime` (see the `loop` function).

Unix time

Unix time or Posix time is defined as the number of seconds that have elapsed since 00:00:00 Coordinated Universal Time (UTC), Thursday, 1 January 1970 (`http://bit.ly/1E6LP3m`).

Main program

Let's start with a simplified version of the main loop:

```
void loop() {

  WiFiClient client;

  client = server.available();
```

```
if (client) {
  Serial.println(F("iOS Device connected"));

  // Waits for client disconnection
  while (client.connected()) {

    // Waits for data and process them

    while (client.available()) {


    }

  }

  Serial.println(F("iOS Device disconnected"));
}


//Serial.println(millis() / 1000 - lastActivation);
if (millis() / 1000 - lastActivation >= ACTIVATION_CHECK_INTERVAL &&
!manualMode) {

  lastActivation = millis() / 1000;
  checkActivations();
}

delay(50);
}
```

Strings in flash memory

Usually Arduino stores static strings in RAM. Since RAM is also used for storing variables, we can move static strings to flash memory with the F() notation. For example, `Serial.println("iOS Device disconnected")`, wastes 23 bytes of RAM for storing the string `"iOS Device disconnected"`. Instead, writing `Serial.println(F("iOS Device disconnected"))` will let the string be stored in flash.

The variable `server` represents the TCP server listening for connections. If a new client is connected and has data available for reading, `available` returns an instance of `WiFiClient` (`client`), which can be used for reading data. While the client is connected, (function `connected` returns true), we check if data is available. The function `available` returns the number of bytes available for reading, so while it returns a number greater than 0, we can read data and process them.

The function `millis` returns the number of milliseconds since the board has been turned on, with the following instructions:

```
if (millis() / 1000 - lastActivation >= ACTIVATION_CHECK_INTERVAL &&
!manualMode) {

    lastActivation = millis() / 1000;
    checkActivations();
}
```

The function `checkActivations` is called every `ACTIVATION_CHECK_INTERVAL` seconds.

The `checkActivations` function checks each activation (which has a name) and based on the current time, turns the appliance on or off.

If the current time is between the activation start time and activation start time plus the activation length, the appliance has to be powered up.

Please, note that the start time is in seconds since 1 January 1970. On the other hand, the length is in minutes.

If the appliance is turned on and the current time is greater than the start time plus the activation length, it has to be turned off. Then the activation is scaled of one period in order to be ready for the next activation.

```
activations[i].startTime += 60 * activations[i].period;
```

 Since `checkActivations` is called every `ACTIVATION_CHECK_INTERVAL` seconds, an activation may be delayed by `ACTIVATION_CHECK_INTERVAL` with respect to its original time.

Now we can take a look at the commands received from the iOS device (refer to the downloaded code).

Each command starts with a byte, which represents the code of the command, and is followed by any additional data.

If the first byte is:

- `'A'`: Arduino sends all the activations to the iOS devices, using the function `sendActivations`.

- `'U'`: It's followed by one byte that represents the index of the activation (idx) to update, and by `sizeof(activation)` bytes, which represent the activation to update. Those bytes are copied on the existing activation with `memcpy((uint8_t *)&activations[idx], (uint8_t *)&inBuffer[2], sizeof(activation))`.

- `'D'`: It's followed by one byte that represents the index of the activation (idx) to delete.

- `'S'`: It's followed by one byte that represents the new state of the appliance which is set with `digitalWrite(PHOTOISOLATOR_PIN, HIGH)` or `digitalWrite(PHOTOISOLATOR_PIN, LOW)`. If the user forces on the state of the appliance, the program enters in manual mode and activations are ignored.

Moreover, until the iOS device is connected, Arduino sends its status using the function `sendStatus`, the first byte is the operation mode (manual or automatic) and the second one is the status of appliance (turned on or off).

The last function we have to look at is `updateActivations`. Let's suppose that we have set an activation at 1:00 P.M. which lasts for 1 minute and repeats every 2 minutes. At 12:59 P.M. we turn off the Arduino and then restart it at 2 P.M. Since 2 P.M. > 1:02 P.M. (now > `startTime` + 60 *period) the activation doesn't start anymore. `updateActivations` has exactly the purpose to shift in time the repeating activation, so that they can be fired properly.

`updateActivations` is then called from the setup function and when the user gets back to automatic operations because during manual operations the activations are not checked and properly updated.

Using function `millis` to calculate current time may lead to significant errors after few days, so we periodically update time via the NTP Server with:

```
// Time synchronization
if (millis() / 1000 > TIME_SYNC_INTERVAL && !updatingTime) {

  updatingTime = true;
  askTime();
}
```

The predefined value for `TIME_SYNC_INTERVAL` updates the current time every 24 hours.

iOS code

In this chapter, we are going to look at the iOS application to manually turn on and off the electrical appliance and managing the activations for the automatic operations.

The full code of this project can be downloaded from here:

`https://www.packtpub.com/books/content/support`

To better understand the explanations in the next paragraphs, you should open the downloaded code while reading.

Creating the Xcode project

The first step is to create a new project. We are going to use the template Tabbed Application again, because it provides two view controllers. In this project, we will also add another one.

Let's create a new Tabbed Application project exactly as you did in the previous chapter and name it `PowerPlug`. Then:

1. Rename `FirstViewController` as `PowerPlugViewController`.
2. Rename `SecondViewController` as `WiFiConnectionViewController`.

In this chapter we are going to use an additional library (CocoaAsyncSocket see http://bit.ly/1NGHDHE), which simplifies the communication over a TCP/IP socket. To install the library you can follow these steps:

1. Open the URL http://bit.ly/1NGHDHE.

2. Click on the **Download ZIP** button (right side of the page). The file CocoaAsyncSocket-master.zip is downloaded into the Downloads folder.

3. Unzip the downloaded file.

4. Locate GCDAsyncSocket.h and GCDAsyncSocket.m.

5. Drag these files and drop them to the Xcode project into the **PowerPlug** group.

6. Be sure that this option is set **Copy Items if needed** (see the following screenshot) and click **Next**.

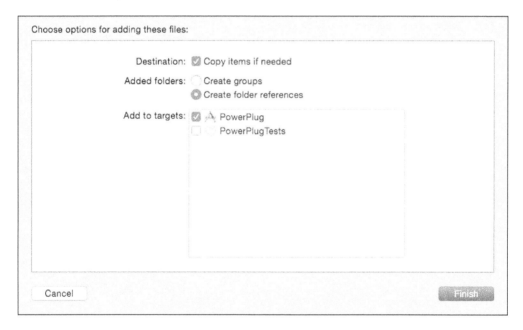

Adding a new view controller

We now add the additional view controller class and the additional view controller graphical container that we need, using the following steps:

1. Select the `PowerPlug` folder in the left panel and right click on it.

2. Select **New File…**.

3. In the left pane select **iOS Source** and in the right one select **Cocoa Touch Class**, then click on **Next**.

4. In the **Subclass** of the list box, select **UITableViewController**.

5. In the **Class** field enter `ActivationsTableViewController` (refer to the next screenshot) and click on **Next**.

6. Click on **Save** on the next window.

7. Select **Main.storyboard** in the left panel.

8. Open the Utilities panel (**View | Utilities | Show Utilities**).

9. In the Utilities panel's search field enter `UIViewController`.

10. Drag the UIViewController to the storyboard.

11. Select the just added view controller.

12. Open the Identity Inspector (**View | Utilities | Show Identity inspector**, or click on the icon circled in red in the next picture).

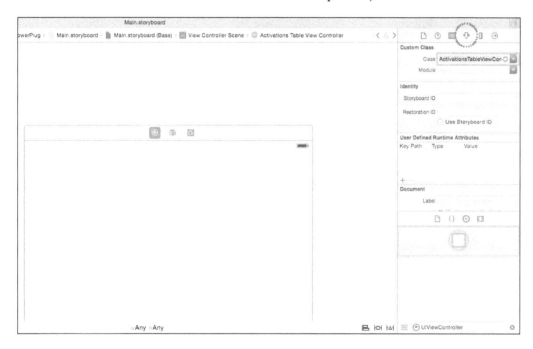

13. In the **Class** list box, select **ActivationsTableViewController**. View controller's class and the GUI are now tied together.

14. We now have to add the new view controller into the Tab Bar of the main view controller. Pressing the *Control* key, drag your mouse pointer from the **Tab Bar Controller** to the **ActivationsTableViewController** and then release (see the next screenshot).

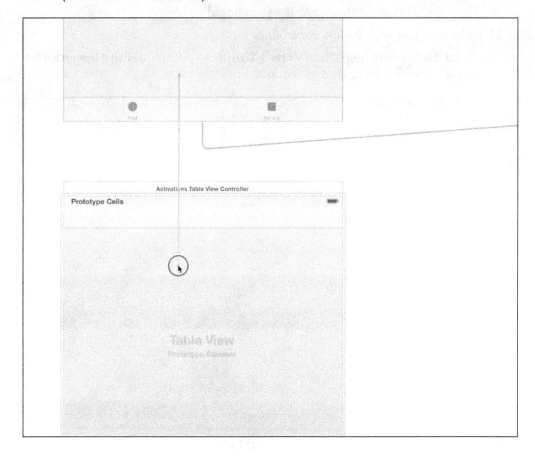

15. When the small dialog shown in the next screenshot appears, select **Relationship Segues | view controllers**.

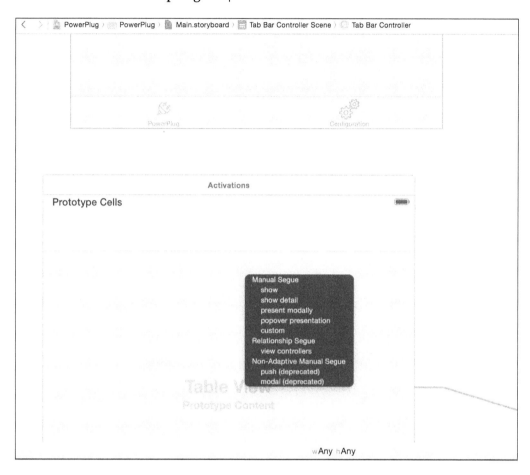

Now the new view controller is added to the **Tab Bar** (see next screenshot).

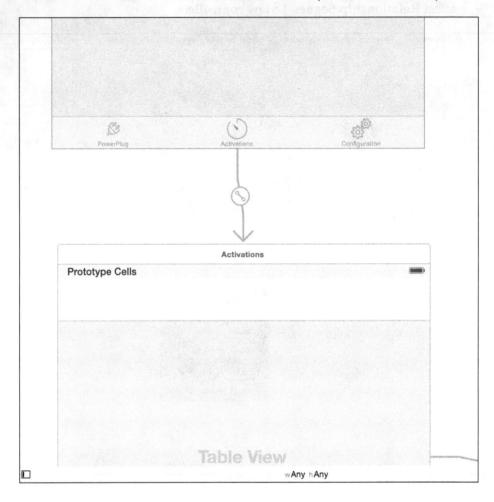

We need the new view control to be the second one in the Tab Bar. To move it, you can click and drag it to the desired position.

Don't forget to change the text in the Tool Bar, respectively to `PowerPlug`, `Activations`, and `Configuration`.

> In the downloaded code, you find three icons to use with each view controller.

Since we need to show details of the elements in the table, we need to embed the ActivationsTableViewController in a navigation controller:

1. Select **ActivationsTableViewController**.

2. Select **Editor | Embed in... | Navigation Controller**.

You should end up with a structure similar to the one shown in the next screenshot:

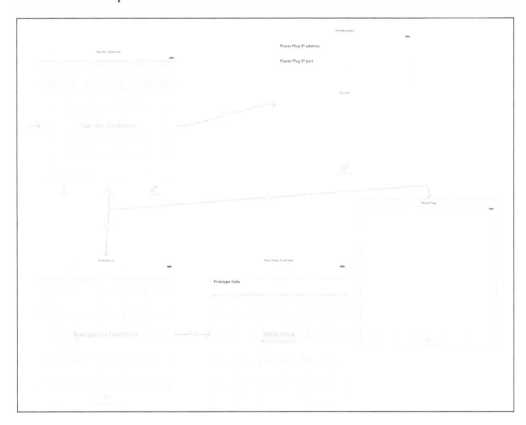

Adding a class for storing the information of each activation

To store information about each activation, we need a class `Activation`.

Model View Controller

Technically speaking, we are now creating the "Model" of the Model View Controller pattern. You can find a brief introduction of the MVC pattern here: `http://apple.co/1hkUDbU` and a complete discussion here: `http://apple.co/1EEpNzL`.

To create the class:

1. Select the **PowerPlug** group in the left panel, right-click it and then select **New File…**.
2. In the left panel, select **Source** and in the right panel, select **Cocoa Touch Class**, then click on **Next**.
3. Select **NSObject** in the **Subclass** list box.
4. Enter `Activation` in **Class Text Field** (see next picture) and click **Next**.
5. Click on **Save**.

6. Open the `Activation.h` file and enter the following code:

```
@interface Activation : NSObject

@property (nonatomic,strong) NSString    *name;
@property (nonatomic,strong) NSDate      *start;
```

```
@property            NSInteger  length;   // minutes
@property            NSInteger  period;   // minutes

@end
```

The project is now ready for the next steps, and we can start working on the view controllers.

Designing the application user interface for WiFiConnectionViewController

As we did in the previous project, we start from the view controller, which allows us to enter the connection information.

It is made of two labels, two fields to enter the IP address, the IP port assigned to the Arduino and a button to update the connection information.

Please refer to the following screenshot to design it:

Since both fields accept only numbers and periods, we can make a properly set keyboard to help the user. To do this:

1. Select a field.
2. Open the Identity Inspector (**View** | **Utilities** | **Show Identity inspector**).
3. Select **Numbers and Punctuation** in the **Keyboard Type** list box.

Now, link the user interface components to the code ending with the following in `WiFiConnectionViewController.m`:

```
@interface WiFiConnectionViewController ()

@property (strong, nonatomic) IBOutlet UITextField *ipField;
@property (strong, nonatomic) IBOutlet UITextField *portField;

@end
```

and:

```
- (IBAction)updateConnectionInformation:(UIButton *)sender {

}
```

Don't forget to link the `UITextFields` delegate to the view controller.

Refer to the downloaded code for more details and compare your results with the provided application.

Designing the application user interface for PowerPlugViewController

This view controller manages the manual operation with the power plug that is capable of manually switching on and off the power to the hooked device.

The final layout of this view controller is shown in the next screenshot:

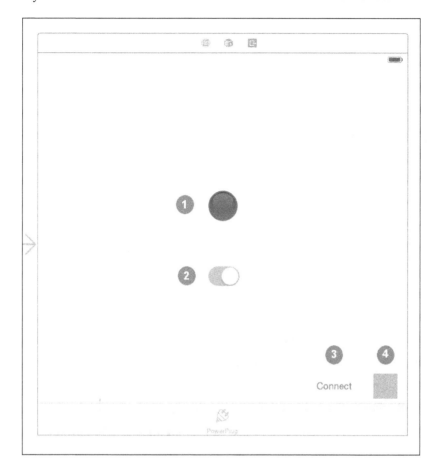

The component numbered **1** is a UIImageView and its purpose is to show if the hooked appliance is powered on or off. To add this component, just drag and drop it and select the image (LEDdisabled.png that you can find in the downloaded project) to to be displayed, using the **Attributes Inspector**. The size of the image view is 60 × 60, which can be set adding the proper layout constraints.

The component numbered **2** is a switch button (`UISwitch`) to actually switch the appliance on and off manually. We can add this component to the container like we did with the others.

The component numbered **3** is a button (`UIButton`) to reconnect to Arduino in case the connection is lost and the component numbered **4** is a `UIView`, which shows if the Arduino is connected or not by the means of its color (light gray: disconnected, green: connected).

Once you have added the components and the required Auto Layout constraints, you can link the components to the code ending with:

```
@interface PowerPlugViewController ()

@property (strong, nonatomic) IBOutlet UIImageView  *applianceStatus;
@property (strong, nonatomic) IBOutlet UIView       *connectionStatus;
@property (strong, nonatomic) IBOutlet UISwitch
*manualOperationButton;

@end
```

Designing the application user interface for ActivationsTableViewController

The ActivationsTableViewController view controller shows all the existing activations in a table.

First, we have to add a class to store information about the cell of each row in the table. To add this class, follow the same steps used in the previous chapters subclassing from `UITableViewCell` and calling the class `ActivationTableViewCell` (see next screenshot).

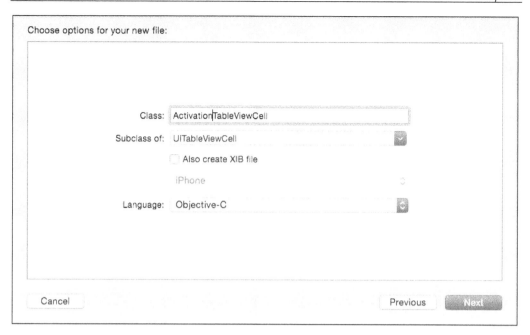

In the `ActivationTableViewCell.h` file, change the code as follows:

```
@interface ActivationTableViewCell : UITableViewCell

@property (strong, nonatomic) IBOutlet UILabel *name;
@property (strong, nonatomic) IBOutlet UILabel *start;
@property (strong, nonatomic) IBOutlet UILabel *end;
@property (strong, nonatomic) IBOutlet UILabel *period;


@end
```

We are now ready to create the custom cell to show each activation information of the power plug, and link it with the ActivationTableViewCell. We are going to show only the main steps, but you can refer to the downloaded code for all the details.

1. In the **Main.storyboard**, select the **ActivationsTableViewController** and then the **Prototypes Cells** in it.

2. Open the Identity Inspector and select **ActivationTableViewCell** in the **Class** list box.

3. Select the Attribute Inspector.

4. Enter `activationCell` in the **Identifier** field.

5. Select the Size Inspector and 66 in the **Row Height** field.

6. Now you can enter the labels (UILabel) to show the different information of each activation, change colors and font sizes, and add the layout constraints ending with a cell prototype like that shown in the next screenshot:

We now have to link the cell components to the class ActivationTableViewCell.

1. Go to **Editor | Show Document Outline** (you can also open this panel using the icon circled red in the next screenshot).

2. Locate the **Activations Table View Controller Scene** and expand it (see next screenshot).

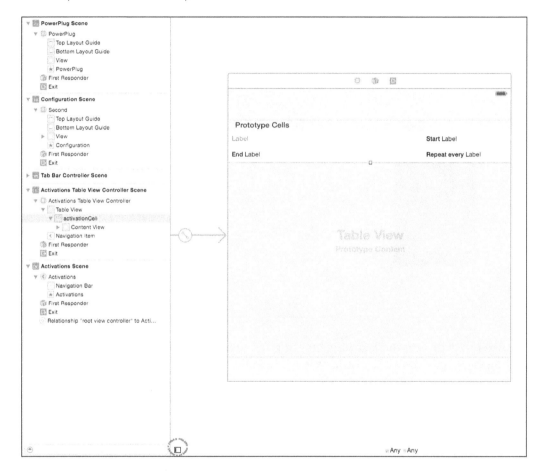

3. Select the **activationCell** and right-click on it.

4. Keeping the *Control* key pressed, drag each property of the class to the related graphical component (see the next screenshot).

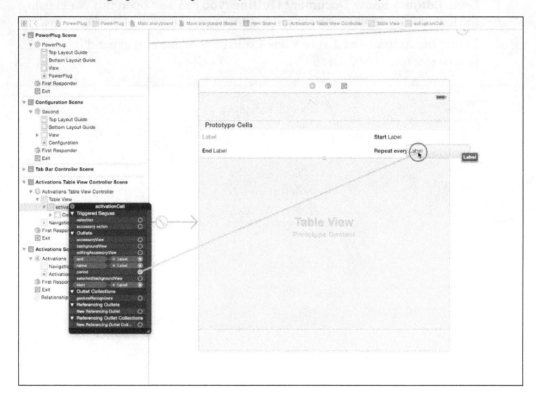

We need to complete the last step: creating the view controller for entering and editing each activation (ActivationViewController).

We start adding a new class inherited from UIViewController called `ActivationViewController`, then we drag and drop a new view controller into the **Main.Storyboard**, and we change its class to `ActivationViewController` in the **Identity Inspector**.

This view controller uses two new components:

- UIDatePicker: It allows the user to choose a date. In our case, the date at which the activation has to start.

- UISegmentControl: It allows the user to easily choose the length of activation and how often the activation has to be started.

The final layout of the controller is shown in the following screenshot:

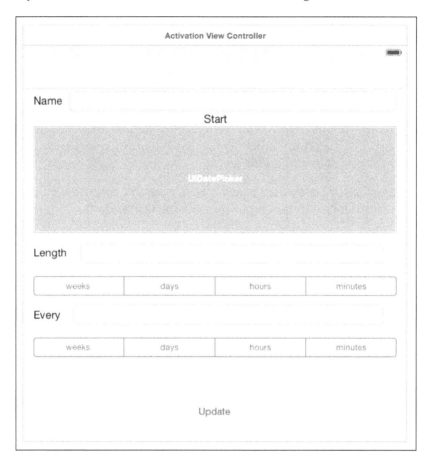

The component can be dropped into the container as we already have learned, and we can set their layout constraints as we did in the other view controllers. The components are linked to the code as usual:

```
@property (nonatomic,weak) Activation *activation;

@property (strong, nonatomic) IBOutlet UITextField
*nameField;
@property (strong, nonatomic) IBOutlet UIDatePicker        *date;
```

```
@property (strong, nonatomic) IBOutlet UITextField          *length;
@property (strong, nonatomic) IBOutlet UISegmentedControl
*lengthScale;
@property (strong, nonatomic) IBOutlet UITextField          *period;
@property (strong, nonatomic) IBOutlet UISegmentedControl
*periodScale;

@end
```

In order to edit activations, we need the ActivationViewController to show when the user taps on a row of the table. To do this, we have to create a segue from the cell to the view controller with the following steps:

1. Keeping *Control* pressed, drag your mouse pointer from the cell to the ActivationViewController (see the following screenshot).

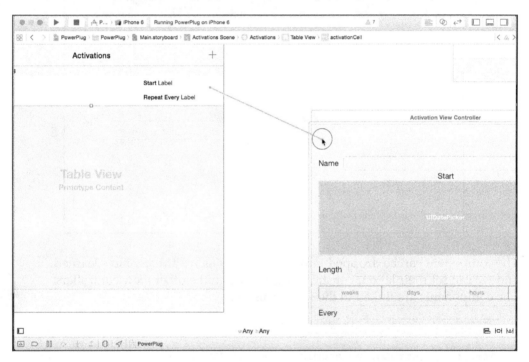

2. When the black dialog appears, select **show**:

You can find all the details about the ActivationViewController in the downloaded code.

This ActivationsTableViewController is finally completed. Let's take a breath and sip a cup of coffee before getting into writing the code of our view controllers. It has been a long and complex run, and we actually deserve a cup of coffee (or tea if you prefer!).

Writing code for the WiFiConnectionViewController

This view controller has the purpose to make sure the user enters the IP address and IP port assigned to Arduino in order to connect to it.

In this project, we are going to store this information in a file, instead of storing it in User Preferences as we did in the Pet Door Locker.

When the update button is tapped, this code is executed:

```
- (IBAction)updateConnectionInformation:(UIButton *)sender {

    if (![self validateIpAddress:_ipField.text]) {


        ...

        return;
    }

    if ([_portField.text integerValue]<0 || [_portField.text
integerValue]>65535) {


        ...

        return;
    }

    NSMutableDictionary *connectionInformation = [[NSMutableDictionary
alloc] init];

    [connectionInformation setValue:_ipField.text forKey:@"IP"];
    [connectionInformation setValue:_portField.text forKey:@"PORT"];

    NSArray *paths = NSSearchPathForDirectoriesInDomains(NSDocumentDir
ectory, NSUserDomainMask, YES);
    NSString *documentsDirectory = [paths objectAtIndex:0];
    NSString *path = [documentsDirectory stringByAppendingPathComponen
t:@"connection.plist"];

    [connectionInformation writeToFile:path atomically:YES];
}
```

The values of the two fields are stored in an NSMutableDictionary (connectionInformation) and stored into a file called writeToFile. This file is saved into a directory documentsDirectory that is accessible by the application only.

Before saving the information, we need to check that the IP address is properly formatted, and that the port is in the allowed range (see the downloaded code for all the details).

Since we set the text field's delegate properties, each time the user is done editing the field, one of the following methods is called and we can make the keyboard disappear using `resignFirstResponder`.

```
- (void)textFieldDidEndEditing:(UITextField *)textField {

    [textField resignFirstResponder];

}

- (BOOL)textFieldShouldReturn:(UITextField *)textField {

    [textField resignFirstResponder];

    return YES;
}
```

When we tap on the text field Period, the keyboard covers it (at least on smaller devices like the iPhone). To avoid this issue, we can shift up the entire view, using a couple of delegate methods of UITextView. When we tap the field, the `textFieldDidBeginEditing` method is called, and when we exit the field or tap return, the `textFieldDidBeginEditing` method is called.

In the `textFieldDidBeginEditing` method we can translate the field, assigning a translation transformation to the field with:

```
self.view.transform = CGAffineTransformMakeTranslation(0, -100);
```

Then we end with these two additional methods:

```
- (void)textFieldDidBeginEditing:(UITextField *)textField {

    if ([textField isEqual:_period]) {

        self.view.transform = CGAffineTransformMakeTranslation(0,
    -100);
    }

}
```

```
- (void)textFieldDidEndEditing:(UITextField *)textField {

    if ([textField isEqual:_period]) {

        self.view.transform = CGAffineTransformMakeTranslation(0, 0);
    }
}
```

Each time this view controller is shown, both fields are filled in with the existing values.

```
- (void)viewDidLoad {
    [super viewDidLoad];


    NSArray *paths = NSSearchPathForDirectoriesInDomains(NSDocumentDir
ectory, NSUserDomainMask, YES);
    NSString *documentsDirectory = [paths objectAtIndex:0];
    NSString *path = [documentsDirectory stringByAppendingPathComponen
t:@"connection.plist"];

    NSDictionary *connectionInformation = [NSDictionary dictionaryWith
ContentsOfFile:path];

    if (connectionInformation != nil) {
        _ipField.text = [connectionInformation objectForKey:@"IP"];
        _portField.text = [connectionInformation
objectForKey:@"PORT"];
    }

}
```

Writing code for AppDelegate

In this project, the connection to Arduino and the communication is managed into `AppDelegate`.

For this purpose, we need some properties:

```
@interface AppDelegate ()

@property (strong, nonatomic) GCDAsyncSocket
*socket;
```

```
@property (strong, nonatomic) NSMutableArray
*activations;

@property (strong, nonatomic) PowerPlugViewController
*powerPlugViewController;
@property (strong, nonatomic) ActivationsTableViewController
*activationsViewController;

@end
```

The `socket` property (provided by the library we added) is a channel used for sending and receiving data from Arduino. The array `activations` stores all the activations the user creates.

When the app starts or enters in the foreground, it starts a connection with Arduino:

```
-(void)connect {

    NSError *err = nil;

    NSArray *paths = NSSearchPathForDirectoriesInDomains(NSDocumentDir
ectory, NSUserDomainMask, YES);
    NSString *documentsDirectory = [paths objectAtIndex:0];
    NSString *path = [documentsDirectory stringByAppendingPathComponen
t:@"connection.plist"];

    NSDictionary *connectionInformation = [NSDictionary dictionaryWith
ContentsOfFile:path];

    if (![_socket connectToHost:[connectionInformation
objectForKey:@"IP"]
                          onPort:[[connectionInformation
objectForKey:@"PORT"] integerValue]
                           error:&err]) {
        NSLog(@"Connection Failed %@", [err localizedDescription]);

        return;
    }

    [_socket readDataWithTimeout:5 tag:0];
}
```

If the connection to Arduino is successful, the delegate method `didConnectToHost` is called.

The method `readDataWithTimeout` allows you to receive data from the counterpart: when data is available, the method `didReadData` is called.

Since we have set a timeout (5 seconds), if no data is received within this time interval, the method `socketDidDisconnect` is called.

Let's take a look at each method.

```
- (void) socket: (GCDAsyncSocket *) sender didConnectToHost: (NSString *)
host port: (UInt16) port {

    [_powerPlugViewController arduinoConnected];
}
```

Here, we just notify the `PowerPlugViewController` instance that the connection was successful.

```
- (void) socketDidDisconnect: (GCDAsyncSocket *) sock withError: (NSError
*) error {

    [_powerPlugViewController arduinoDisconnected];
    _activations = nil;
}
```

Here, we just notify the `PowerPlugViewController` instance that the Arduino has disconnected, and we release the activations we were using.

From Arduino we can receive two kinds of messages:

- **Status**: It starts with character `'S'` and it's followed by two bytes:
  - First byte is 1 if the appliance has been turned on manually, 0 otherwise
  - Second byte is the status of the appliance: 0 if it's switched off, 1 if it is switched on

- **Activations**: It starts with character `'A'` and it's followed by `ACTIVATION_SIZE_ON_ARDUINO * NUMBER_OF_ACTIVATIONS` bytes which represent activations

Messages coming from Arduino are received in the following method, and are processed following the previous rules.

```
- (void)socket:(GCDAsyncSocket *)sender didReadData:(NSData *)data
withTag:(long)tag {

    NSLog(@"Bytes received %lu",(unsigned long)[data length]);

    if ([data length]<3) {
        return;
    }

    NSString* answerType = [[NSString alloc] initWithData:[data
subdataWithRange:NSMakeRange(0, 1)]
                                            encoding:NSASCIIStringEn
coding];

    // S<manual><appliance status>
    // A<activations>


    if ([answerType isEqualToString:@"S"]) {

        BOOL manual;
        BOOL status;

        [[data subdataWithRange:NSMakeRange(1, 1)] getBytes: &manual
length: 1];
        [[data subdataWithRange:NSMakeRange(2, 1)] getBytes: &status
length: 1];

        [_powerPlugViewController updateStatus:manual
applianceStatus:status];
    }

    if ([answerType isEqualToString:@"A"]) {

        if (data.length < (ACTIVATION_SIZE_ON_ARDUINO * NUMBER_OF_
ACTIVATIONS + 1)) {

            NSLog(@"Error reading data");
```

```
                    return;
            }

        for (int i=0; i<NUMBER_OF_ACTIVATIONS; i++) {

            NSData *nameData = [data subdataWithRange:NSMakeRange(1+AC
TIVATION_SIZE_ON_ARDUINO*i, 21)];
            NSString *name = [NSString stringWithCString:nameData.
bytes encoding:NSASCIIStringEncoding];

            if (name.length>0) {

                NSData *startData = [data subdataWithRange:NSMakeRange
(1+ACTIVATION_SIZE_ON_ARDUINO*i+21, 4)];

                NSInteger start=0;
                [startData getBytes: &start length: 4];

                NSData *lengthData = [data subdataWithRange:NSMakeRang
e(1+ACTIVATION_SIZE_ON_ARDUINO*i+21+4, 2)];

                NSInteger length=0;
                [lengthData getBytes: &length length: 2];

                NSData *periodData = [data subdataWithRange:NSMakeRang
e(1+ACTIVATION_SIZE_ON_ARDUINO*i+21+4+2, 2)];

                NSInteger period=0;
                [periodData getBytes: &period length: 2];

                Activation *activation = [[Activation alloc] init];
                activation.name = name;
                activation.start = [NSDate dateWithTimeIntervalSince1
970:start];
                activation.length = length;
                activation.period = period;

                [_activations addObject:activation];
            }
        }

        [_activationsViewController dataReceived];
    }
```

```
    [sender readDataWithTimeout:5 tag:0];
}
```

If a Status message is received, the `PowerPlugViewController` instance is notified:

```
[_powerPlugViewController updateStatus:manual
applianceStatus:status];
```

If an Activations message is received, each activation is added to `_activations`:

```
[_activations addObject:activation];
```

In the end, the instance of ActivationsViewController is notified that all the activations are available:

```
[_activationsViewController dataReceived];
```

AppDelegate also implements delegate methods for PowerPlugViewController and ActivationsViewController view controllers.

For ActivationsViewController, the implemented methods are `getActivations`, `updateActivationOfIndex`, and `deleteActivationOfIndex`.

The method `getActivations` is called by the ActivationsViewController when it needs to show the list of configured activations.

```
-(NSMutableArray *)getActivations {

    if (_activations == nil) {

        _activations = [[NSMutableArray alloc] init];

        NSString *message = @"A";
        [_socket writeData:[message dataUsingEncoding:NSASCIIStringEnc
oding] withTimeout:-1 tag:0];

        [_activationsViewController dataRequested];

        return _activations;
    }

    return _activations;
}
```

If `_activations` is nil, a message to Arduino is sent which will respond with the list of all the activations. The message is very easy: only one byte with the value `'A'` (Activations).

The message is actually sent to Arduino calling:

```
[_socket writeData:[message dataUsingEncoding:NSASCIIStringEncoding]
withTimeout:-1 tag:0];
```

The method `updateActivationOfIndex` is called by ActivationsViewController when an activation has been updated (or added). The method `deleteActivationOfIndex` is called by ActivationsViewController when an activation has to be deleted.

For the complete implementation of these two methods, look at the downloaded code.

Methods implemented for PowerPlugViewController are `sendSwitchCommand` and `reconnect`, which are very easy and you can understand them directly from the downloaded code.

Writing code for PowerPlugViewController

This view controller manages manual operations of the appliance attached to the power plug.

First, we need to declare the delegate methods and messages to which this view controller responds.

```
@protocol PowerPlugViewControllerDelegate <NSObject>

-(void)sendSwitchCommand:(BOOL)on;
-(void)reconnect;

@end

@interface PowerPlugViewController : UIViewController

-(void)arduinoConnected;
-(void)arduinoDisconnected;
-(void)updateStatus:(BOOL)manual applianceStatus:(BOOL)
applianceStatus;
```

```
@property (strong, nonatomic) id<PowerPlugViewControllerDelegate>
delegate;

@end
```

This view controller is very easy and its methods are quite auto explicative and self explanatory. We just need to take a look at `updateStatus`:

```
-(void)updateStatus:(BOOL)manual applianceStatus:(BOOL)applianceStatus
{

    _manualOperationButton.on = manual;

    if (applianceStatus)
        _applianceStatus.image = [UIImage imageNamed:@"LEDon.png"];
    else
        _applianceStatus.image = [UIImage imageNamed:@"LEDoff.png"];
}
```

When the `applianceStatus` is true, the appliance is turned on, and we set the image `LEDon.png` to the imageView (`_applianceStatus`). This image simulates an LED turned on. When the appliance is turned off, the image shown is `LEDoff.png`, which simulates an LED being turned off.

Writing code for ActivationsTableViewController

This view controller manages the list of activations, allowing the user to add, delete, and update each activation. It is based on the UITableView component, which is one of the most used and powerful components of the UIKit.

Let's start from `ActivationsTableViewController.h`:

```
@protocol ActivationsTableViewControllerDelegate <NSObject>

-(NSMutableArray *)getActivations;
-(void)updateActivationOfIndex:(uint8_t)index;
-(void)deleteActivationOfIndex:(uint8_t)index;

@end

@interface ActivationsTableViewController : UITableViewController
<ActivationViewControllerDelegate>
```

```
@property (strong, nonatomic) id<ActivationsTableViewControllerDelega
te> delegate;

-(void)dataRequested;
-(void)dataReceived;

@end
```

By now, you should be able to recognize the delegate protocol and messages to which the view controller responds.

Let's start from the implementation of the UITableView delegate methods. The first is numberOfRowsInSection, which is called from the table when it needs to know how many items it has to show:

```
- (NSInteger)tableView:(UITableView *)tableView numberOfRowsInSection:
(NSInteger)section {

    return [_delegate getActivations].count;
}
```

It doesn't require many explanations. The method cellForRowAtIndexPath is called by the UIViewTable for each row to be shown:

```
- (UITableViewCell *)tableView:(UITableView *)tableView cellForRowAtIn
dexPath:(NSIndexPath *)indexPath {

    ActivationTableViewCell *cell = [tableView dequeueReusableCellWith
Identifier:@"activationCell"];
    if (cell == nil) {
        cell = [[ActivationTableViewCell alloc] initWithStyle:UITableV
iewCellStyleDefault reuseIdentifier:@"activationCell"];
    }

    Activation *activation = [_delegate getActivations][indexPath.
row];

    cell.name.text = activation.name;

    NSDateFormatter *dateFormatter =  [[NSDateFormatter alloc] init];
    [dateFormatter setDateStyle:NSDateFormatterShortStyle];
    [dateFormatter setTimeStyle:NSDateFormatterMediumStyle];
```

```
    NSDate *endDate = [activation.start dateByAddingTimeInterval:60*ac
tivation.length];
    cell.start.text = [dateFormatter stringFromDate:activation.start];
    cell.end.text = [dateFormatter stringFromDate:endDate];


    ....


    return cell;
}
```

Calling the method [tableView dequeueReusableCellWithIdentifier:@"activa tionCell"] we get a cell and we fill it with the values of one activation.

If the previous function returns nil, there is not an available cell, and one has to be created with:

```
cell = [[ActivationTableViewCell alloc] initWithStyle:UITableViewCellS
tyleDefault reuseIdentifier:@"activationCell"];
```

Take a look at the downloaded code to see the full method implementation.

When a row is deleted from the UITableView, the following method is called:

```
- (void)tableView:(UITableView *)tableView commitEditingStyle:(UITabl
eViewCellEditingStyle)editingStyle forRowAtIndexPath:(NSIndexPath *)
indexPath {


    if (editingStyle == UITableViewCellEditingStyleDelete) {

        [[_delegate getActivations] removeObjectAtIndex:indexPath.
row];

        [tableView deleteRowsAtIndexPaths:[NSArray
arrayWithObject:indexPath] withRowAnimation:UITableViewRowAnimationFa
de];

        [_delegate deleteActivationOfIndex:indexPath.row];
    }
}
```

Method `deleteRowsAtIndexPaths` removes the row from the table and the delegate method `deleteActivationOfIndex` creates a message that is then sent to Arduino for deleting an activation.

When designing the interface we created a segue from the table cell to the ActivationViewController in order that when the user taps a table row, the ActivationViewController appears.

Before the ActivationViewController is started, the method `prepareForSegue` is called:

```
- (void)prepareForSegue:(UIStoryboardSegue *)segue sender:(id)sender {

    ActivationViewController *activationViewController =
(ActivationViewController *)[segue destinationViewController];
    activationViewController.delegate = self;

    NSIndexPath *p = [self.tableView indexPathForSelectedRow];

    NSMutableArray *activations = [_delegate getActivations];
    if (activations==nil)
        return;

    _selectedActivationIndex = p.row;

    [self.tableView deselectRowAtIndexPath:p animated:NO];
}
```

Here we set the delegate property of ActivationViewController and we store the index of the selected row.

Segue identifier

The method `prepareForSegue` is called for each segue that starts from the view controller. Usually, different target view controllers require different initialization code. In the Interface Builder you can set the **Identifier** of each segue, and use these instructions for distinguishing between segues:

```
if ([segue.identifier isEqualToString:"<identifi
er>"]) {

    ....

}
```

The ActivationViewController requires two delegate methods, which don't require any explanation:

The first is `getActivation`:

```
-(Activation *)getActivation {

    return [_delegate getActivations][_selectedActivationIndex];
}
```

The second is `update`:

```
-(void)update {

    [self.delegate updateActivationOfIndex:_selectedActivationIndex];
    [self.tableView reloadData];
}
```

We just point out that `[self.tableView reloadData]` forces the table view to reload data, starting from calling `numberOfRowsInSection` and then calling `cellForRowAtIndexPath` for each row.

Writing code for ActivationTableViewController

This view controller doesn't do much more than getting the user entered values and updating the selected activation.

To understand the details, refer to the code provided with the book.

Testing and tuning

When you have completed your application, you can test the system following these steps:

1. Change IP information (IP, gateway, and so on) and network information (SSID, pass) in the Arduino sketch to adapt them to your own network configuration. You may need to access your router configuration page to get this information.
2. Upload the sketch to Arduino.
3. Check the Arduino console for any error messages.

4. Connect the driver circuit to the power line and to an external appliance (a lamp can do the job). Please, follow all the required security measures to avoid any electric shocks.

5. Run the iOS application on your device or in the simulator.

6. Tap on Configuration and enter the IP address and port you have set in the Arduino sketch.

7. Tap on the power plug, connect and switch the button on and off.
You should see the external appliance turning on and off accordingly.
Then turn off your appliance.

8. Tap on Activations and enter an activation. You should see your appliance automatically turning on and then off accordingly with the time and values you have entered.

If you are not able to get the time from the NTP server, you may try to change the address. To find a new address, you can look at the link `http://tf.nist.gov/ tf-cgi/servers.cgi` or send a `ping` to the address `time.nist.gov` and use the returning address.

How to access the power plug from anywhere in the world

In this section, you will learn the basics to access your Arduino board from outside your home network. In other words, you will be able to use the iOS application through the mobile network to access your power plug behind your home router/firewall.

This section is provided for reference only because there are so many routers available in the market, and there are so many network configurations, that it is actually impossible to provide a complete guide. Anyway, with this brief overview of the matter, you should be able to configure your own devices.

Port forwarding

The next picture shows the typical configuration of a domestic network.

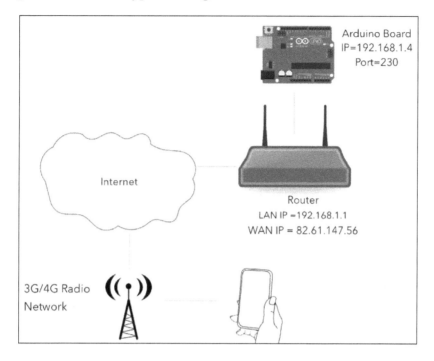

Let's suppose that your IP addresses are:

- Router WAN IP: 82.61.147.56
- Router LAN IP: 192.168.1.1
- Arduino Board IP: 192.168.1.4

Typically, the IP address of the Arduino board on your internal network is not visible from the external network, at least unless you don't explicitly configure your router. This configuration is called IP port forwarding.

Basically, this configuration instructs the router to transfer the traffic it receives on a specific port to another IP address on a specific port.

That said, your port forwarding configuration has to be something like this:

(82.61.147.56, 230) --> (192.168.1.4, 230)

Accessing your router configuration page (usually via browser), you should be able to get the current IP that your Internet service provider has assigned to your router and configure the port forwarding. Read your router manual to learn how to set up this configuration.

Once you have configured port forwarding, you can access the Arduino board using:

- **IP**: 82.61.147.56 (or better, the actual IP address assigned your own router)
- **Port**: 230

Pay attention to security!

Once you enable the port forwarding, anyone can access it. This means that anyone can easily get access to your power plug and control it. We should implement a protocol, which authenticates the iOS device, and encrypts all the messages exchanged. Unfortunately, Arduino does not have enough processing power and memory for this task. If you want to keep communication between an iOS device and Arduino secure, you have to set up a Virtual Private Network (http://bit.ly/1ENoe3o). There are many routers that provide a VPN, and iOS supports it natively.

Dynamic DNS

Typically, each time your router connects to the Internet, it gets a different public IP address. So, you need to change the IP address of Arduino Manager each time your router is restarted. This is not practical.

Many **Dynamic Domain Name Services (DDNS)** are available to dynamically associate the same name to your router IP address, even if the address changes. Most are available for free, among them: NoIP, yDNS, and FreeDNS.

Most routers have DDNS client ready to use and easy to configure, otherwise, you may need to install simple software on a computer connected to your network.

Please visit the service provider's site to sign up and get installation and configuration procedures. Moreover, check your router for DDNS service availability.

Once you have configured the Dynamic DNS service of your choice and supposing that you have chosen a domain like `powerplug.something.com`, you can access the Arduino with:

- **IP**: `powerplug.something.com`
- **Port**: 230

And you don't need to change them when you reboot your router.

How to go further

The project we have developed can be improved in many ways, some improvements that you can try yourself:

- Control more external appliances using more Arduino pins, and more driver circuits.
- Check and report to the user any conflicts between different activations.

Summary

In this chapter, you have built a device that allows you to control an external device (a lamp, a washing machine, a coffee kettle, and so on) manually or automatically, turning it on and off as you like, and/or at a specific time.

You have learned to make a power circuit without using a relay and manage it from Arduino.

On Arduino, you have learned to write a program that reads and writes on the SD card, uses the Wi-Fi shield to communicate to iOS devices and to an external service (to get the current time), and manages an external power circuit for controlling the external device.

On iOS, you have learned how to design the user interface for a medium complex application also using the UITableView that is one of the most used components to interact with the user. You are now able to understand the MVC pattern and take advantage of it in your own programs. Moreover, you have learned to handle TCP/IP communication via TCP socket.

The next chapter will be about coding Arduino and iOS for controlling a rover robot using the hardware features provided by the iOS devices (accelerometer).

4
iOS Guided Rover

In this chapter, we are going to build software that can be used to control a rover robot. What's new in this? Children have been playing with radio-controlled toys for years now, and sometimes, we still play with such toys!

However, we are going to build a robot that can be controlled by iOS devices in the following three different ways:

- Through manual commands: The two sliders on the iOS control: the rover's steering and the throttle
- By means of the iOS device movements in space: By moving an iOS device left and right, you can control the steering, and by moving the iOS device back and forth, you can control the throttle
- By using voice commands: By saying a few voice commands, an iOS device can control the steering and throttle

Cool, isn't it? Let's quickly get started with this fascinating journey into robotics without forgetting that the techniques that you are going to learn here can be used in many other projects.

The chapter is organized into the following sections:

- iOS guided rover requirements: We will briefly set the project requirements
- Hardware: We will describe the hardware and the electronic circuit that is needed for the project
- The Arduino code: We will write the code for Arduino to control the external appliance and communicate with iOS devices
- The iOS code: We will write code for an iOS device
- How to go further: More ideas will be provided to improve the project and learn more

iOS guided rover requirements

We are going to develop the Arduino and iOS software to do the following:

- Control the direction and speed of a rover, powered by two motors from the iOS device by:
    ◦ Using manual commands
    ◦ Moving the iOS device in space
    ◦ Using voice commands

- Avoid obstacles

- Measure the slope and tilt angles so that the robot doesn't flip over

- Transfer commands and information back and forth between the rover and the iOS application by using the Bluetooth BLE.

Hardware

We will assume that you have already built a rover robot like the ones shown on the following sites:

- `http://bit.ly/1MzmIYs`

- `http://bit.ly/1Jo9qJS`

- `http://bit.ly/1i7oTag`

You can also buy two motors and two wheels (for more information, visit `http://bit.ly/1i7oWTG` and `http://bit.ly/1KeHtdf`) and make the chassis yourself with metal or wood.

You will find everything that you may need on eBay (`http://www.ebay.com`) at an affordable price.

Additional electronic components

In this project, we will need some additional components:

- Sparkfun DC Motor Driver TB6612FBG (`https://www.sparkfun.com/products/9457` or `http://www.adafruit.com/product/2448`)

- Sharp GP2Y0A21YK Distance Measuring Sensor Unit (`https://www.sparkfun.com/products/242`), which can be replaced by similar units with minimum changes in the Arduino code

- Adafruit ADXL345 Triple-Axis Accelerometer (http://www.adafruit.com/products/1231)
- 330 Ω resistor 0.25w
- A red LED

What's an accelerometer?

In this section, we will briefly discuss what an accelerometer is and how to use it.

An accelerometer is a device that measures acceleration along its 3 axes, and returns three voltage signals that are proportional to the acceleration.

The acceleration measured by the accelerometer gives us information about the inclination of the accelerometer with respect to the terrestrial gravitational field.

To understand how acceleration is related to the inclination of the accelerometer, let's examine the situation depicted in the following figures, where we consider only two axes (z and x):

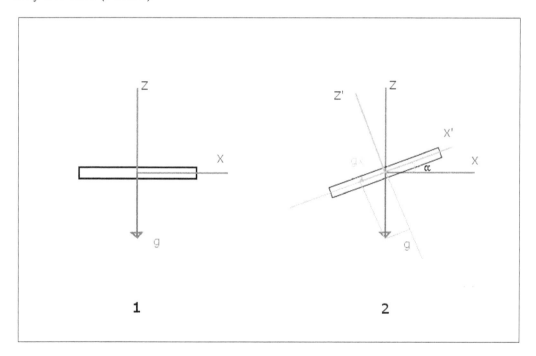

In figure **1**, the accelerometer has the z axis is parallel to the gravity (**g**), and the x axis is orthogonal to the gravity. Hence, the acceleration along the x axis is zero. In figure **2**, the accelerometer rotates at an angle **α**. The acceleration along the accelerometer x axis is gx, and gx is what the accelerometer returns as the acceleration along the x axis. In conclusion, by rotating the accelerometer, we have an acceleration along the accelerometer x axis that is proportional to the rotation itself. The same is true for the y axis that is pointing into the page.

In this project, we use this behavior for mounting an accelerometer on the rover to measure its inclination along the transversal and the longitudinal axis. At the same time, we use the iOS device accelerometer to measure device inclination for controlling the direction and speed of the rover.

Electronic circuit

A DC motor rotates in one direction when powered up and rotates in the opposite direction when the voltage polarity is inverted.

The following circuit diagrams depict ways to use switches to power the DC motor to invert the voltage polarity and rotation direction. In the following circuit diagram, the motor rotates in the clockwise direction:

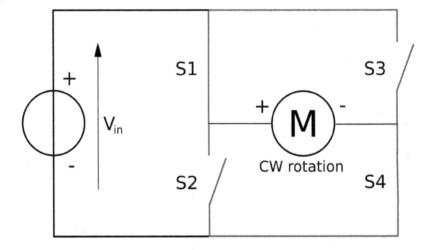

While in the second case, it rotates in the anticlockwise direction:

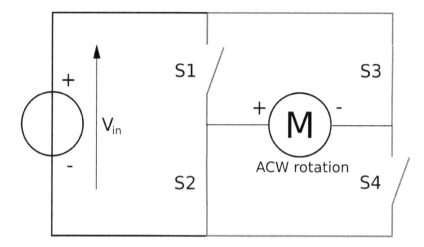

Transistors can replace switches in order to control the direction through electronic signals. A transistor-based circuit that replaces the four switches is called an H-bridge (http://bit.ly/1JBmdrE).

Basically, an H-bridge is a circuit that allows you to control the DC motor direction through two input signals. When the first signal is high and the second one is low, the motor runs in one direction; when the first signal is low and the second one is high, the motor runs in the opposite direction. To control the speed of the motor, we also need a PWM signal (https://www.arduino.cc/en/Tutorial/PWM). The TB6612FBG motor driver includes two H-bridge circuits to power the two motors up to 1A.

The following table, derived from the TB6612FBG datasheet describes how to use the available input signals to control the motor:

Input				Output		
IN1	IN2	PWM	STBY	OUT1	OUT2	Mode
H	H	H/L	H	L	L	Short brake
L	H	H	H	L	H	Motor runs
		L	H	L	L	Short brake
H	L	H	H	H	L	Motor runs in the opposite direction
		L	H	L	L	Short brake
L	L	H	H	OFF (high impedance)		Stop
H/L	H/L	H/L	L	OFF (high impedance)		Standby

To make the motor run in a direction, we have to make IN1 low and IN2 high. The PWM signals control the motor speed. To make the motor run in the opposition direction, we have to make IN1 high and IN2 low, and the PWM signals still control the motor's speed.

The PWM signals can be generated from Arduino by using the analogWrite function.

The following electric diagram of the electronic circuit is what we need for the project:

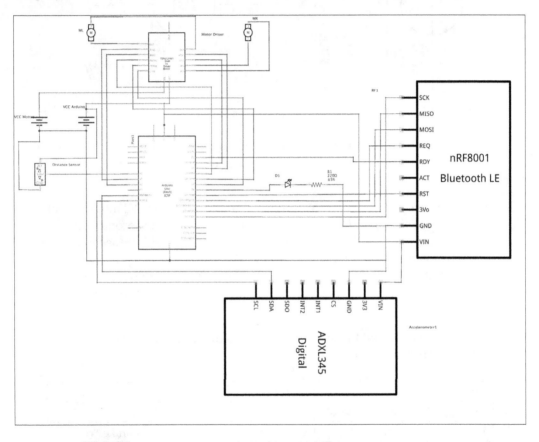

As required, we need a distance sensor that has an analog output; the output voltage decreases as the distance to the front object decreases. The relation between the output voltage and distance is not linear (see the datasheet by visiting `https://www.sparkfun.com/datasheets/Components/GP2Y0A21YK.pdf`), but it can be linearized in the range of the distance of interest. In our application, we just need to stop the rover when it is very close to an obstacle. So, we don't care much about the actual distance. The red LED shows when an obstacle is very close to the rover.

ADXL345 is a 3-axis accelerometer. It measures the acceleration along its three axes (in [m/s^2]). When the rover is completely at rest and leveled, the acceleration is 0 m/s^2 on the x axis, 0 m/s^2 on the y axis, and 9.8 m/s^2 on the z axis because of the gravitational force. When the rover is not leveled, the values read along the axis are different. When the measured values exceed a threshold, we know that the rover is going to flip over.

The following diagram shows how to mount the circuit on a breadboard:

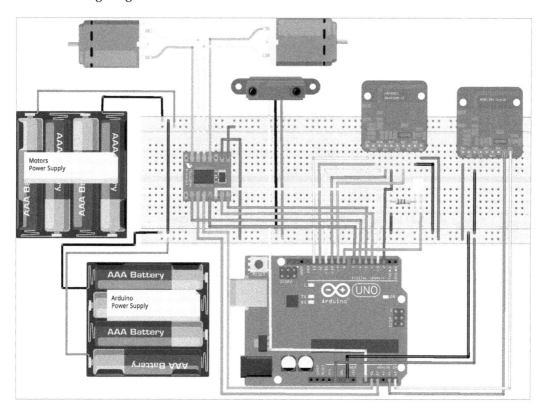

You can see that we used two different power sources, one for Arduino and the electronic components, and the other for the motors. This was done for the following reasons:

1. The motors can require a different voltage compared to the electronic components. The motor power source can have a different voltage up to 15V, as required by the TB6612FBG chip.

2. Motors generate a lot of electric noise that can disturb the electronics. With this configuration, electronics and motors are electrically isolated.

3. The rover can operate for a longer time, especially by using rechargeable batteries.

[

Motor wiring
In case a motor spins in the wrong direction, you have to invert its wires.
]

How to make the rover turn

When both the motors spin at the same speed, the rover goes straight. To make it turn right, reduce the speed of the right motor. The more we reduce the speed, the more the rover turns. This is exactly the same as the method that is used to make the rover turn left.

That said, the Arduino code has to manage the rotation speed of the two motors in order to control the direction of the motor.

How to mount the accelerometer

It's very important that the accelerometer board be mounted in the right position in order to properly read acceleration and work with the Arduino code that is shown in the next chapter. The accelerometer board has to be tightly screwed to the rover chassis with a nut and a bolt (it is better to use two nuts and bolts, one for each side).

The schema is shown in the following image:

Arduino code

The full code of this project can be downloaded from `https://www.packtpub.com/books/content/support`.

To have a better understanding of the explanations in the next few paragraphs, open the downloaded code while reading.

Before going any further, we need to install the following additional libraries from Adafruit:

- **Adafruit ADXL345**: This is used to get measurements from the ADXL345 accelerometer

- **Adafruit Unified Sensor**: This is a general library that is required from the previous one

To install the libraries into the Arduino IDE, open the menu **Sketch | Include Library | Manage Libraries ...**. For more information, see *Chapter 2, Bluetooth Pet Door Locker*.

Setup code

Please refer to the downloaded code, since the setup code is quite simple and doesn't require a detailed explanation.

Just note that currentSpeed, leftSpeed, and rightSpeed are: the current speed of the rover, the speed reduction of the left wheel used to make the rover turn left, and the speed reduction of the right wheel used to make the rover turn right respectively. They are obviously all set to zero in the setup function.

The goingForward variable indicates whether the rover is going forward or backward, and it is initially set to yes (forward).

The final lines initialize the accelerometer, as required by the library. The xOffset and yOffset variables are about tuning the accelerometer readings when the rover is at rest and in a plane. We will talk about them in the *Testing and tuning* section.

Motor control functions

Before getting into the explanation of the main code, we will take a look at the motor control functions, which are as follows:

- forward: This configures the motor control to make the rover move forward
- backward: This configures the motor control to make the rover move backward
- brake: This stops the rover
- throttle: This controls the speed of the rover and its direction

From the previous table for TB6612FBG (row 2), to make the motors go forward, we need to set IN1 to low and IN2 to high. This is exactly what the forward function does for both the motors:

```
void forward() {

  digitalWrite(STBY, HIGH);

  digitalWrite(MR_I1, LOW);
  digitalWrite(MR_I2, HIGH);

  digitalWrite(ML_I1, LOW);
  digitalWrite(ML_I2, HIGH);
}
```

The `backward` function is similar. From row 3 of the previous table, we need to set IN1 to high and IN2 to low to change the direction, as follows:

```
void backward() {

    digitalWrite(STBY, HIGH);

    digitalWrite(MR_I1, HIGH);
    digitalWrite(MR_I2, LOW);

    digitalWrite(ML_I1, HIGH);
    digitalWrite(ML_I2, LOW);
}
```

In both the functions, we set STBY to HIGH in case we previously stopped the rover by setting STBY to LOW.

Again, by using the table (row 5), use the following function to stop the rover:

```
void brake(void) {

    digitalWrite(MR_I1, LOW);
    digitalWrite(MR_I2, LOW);

    digitalWrite(ML_I1, LOW);
    digitalWrite(ML_I2, LOW);

    digitalWrite(STBY, HIGH);
}
```

The `throttle` function is very important as it controls the speed and direction of the rover:

```
void throttle(int requiredSpeed, int requiredLeftSpeed, int
requiredRightSpeed) {

    analogWrite(ML_PWM, requiredSpeed - requiredLeftSpeed);
    analogWrite(MR_PWM, requiredSpeed - requiredRightSpeed);
}
```

To set the speed of each motor, we have to set an appropriate PWM signal to the PWM pin.

If you need to move the rover in a straight line, set the speed in such a way that it is equal for both motors. Otherwise, reduce the speed of the wheel that is on the rover's side, which we want to turn to.

Main program

The `loop` function of the rover control software is not very complex, and is as follows:

```
void loop() {

  uart.pollACI();

  if (iOSConnected) {

    // Check accelerometer
    if (millis() - lastAccelerometerCheck > ACCELEROMETER_CHECK_
INTERVAL) {

      char buffer[32];
      char xBuffer[6];
      char yBuffer[6];

      lastAccelerometerCheck = millis();

      sensors_event_t event;

      accel.getEvent(&event);

      event.acceleration.x += xOffset;
      event.acceleration.y += yOffset;

      dtostrf(event.acceleration.x, 0, 2, xBuffer);
      dtostrf(event.acceleration.y, 0, 2, yBuffer);

      snprintf(buffer, 32, "%s:%s", xBuffer, yBuffer);
      uart.write((uint8_t *)buffer, strlen(buffer));

      //        Serial.print("X: "); Serial.print(event.
acceleration.x); Serial.print("  ");
      //        Serial.print("Y: "); Serial.print(event.
acceleration.y); Serial.print("  ");Serial.println("m/s^2 ");

    }
  }

  // Reads distance
```

```
distance = 0;
for (int i = 0; i < 16; i++)
    distance += analogRead(DISTANCEPIN);
distance = distance / 16;

if (distance > DISTANCETHRESHOLD) {
    leftSpeed = 0;
    rightSpeed = 0;
    throttle(0, 0, 0);
    digitalWrite(DISTANCEINDICATORPIN, HIGH);
}
else {
    digitalWrite(DISTANCEINDICATORPIN, LOW);
}
}
```

When an iOS device is connected, after every ACCELEROMETER_CHECK_INTERVAL milliseconds, the acceleration values along the x and y axes are sent to the iOS device. Then, the distance sensor is read. If the distance from an obstacle is greater than DISTANCETHRESHOLD, the rover stops and the LED on the rover is turned on.

Since the distance sensor readings are quite variable (like with most of the analog sensors), the mean value of 16 readings is used through the following lines:

```
for (int i = 0; i < 16; i++)
    distance += analogRead(DISTANCEPIN);
distance = distance / 16;
```

To control the rover, the iOS function has to send the following commands:

- F: This is used to move the rover forward.
- B: This is used to move the rover backward.
- T=<speed>: This is used to move the rover at a speed of <speed>. The <speed> lies in the 0-100 range.
- R=<speed>: This is used to move the rover right by reducing the current speed of the right motor to <speed>. The <speed> lies in the 0-100 range.
- L=<speed>: This is used to move the rover left by reducing the current speed of the left motor to <speed>. The <speed> lies in the 0-100 range.

As we already know from the *Bluetooth Pet Locker* project in *Chapter 2, Bluetooth Pet Door Locker*, commands from the iOS controller are received in the rxCallback function:

```
void rxCallback(uint8_t *buffer, uint8_t len) {

  if (len > 0) {

    char value[32];

    if (buffer[0] == 'F') {

      forward();
      goingForward = true;
    }

    if (buffer[0] == 'B') {

      backward();
      goingForward = false;
    }

    if (buffer[0] == 'T') {

      strncpy(value, (const char *)&buffer[2], len - 2);
      value[len - 2] = 0;

      currentSpeed = map(atoi(value), 0, 100, 0, 255);
      if (currentSpeed == 0) {
        rightSpeed = 0;
        leftSpeed = 0;
      }
    }

    if (buffer[0] == 'R') {

      strncpy(value, (const char *)&buffer[2], len - 2);
      value[len - 2] = 0;
```

```
        //Serial.print("Right Speed "); Serial.println(atoi(value));
        rightSpeed = map(atoi(value), 0, 100, 0, currentSpeed);
        leftSpeed = 0;
    }

    if (buffer[0] == 'L') {

        strncpy(value, (const char *)&buffer[2], len - 2);
        value[len - 2] = 0;

        leftSpeed = map(atoi(value), 0, 100, 0, currentSpeed);
        rightSpeed = 0;
    }

    throttle(currentSpeed, leftSpeed, rightSpeed);
  }

}
```

The F(orward) and B(ackward) commands are very easy to handle, since we have the respective function to call.

For the T(hrottle) command, we get the required reduction speed in the `value` variable, and we proportionally scale it from the range of 0-100 to the range 0-255 (`currentSpeed = map(atoi(value), 0, 100, 0, 255)`). This is the expected range for the PWM signal.

The function ends by calling the `throttle` function, which sets the speed for both the motors, thus setting the PWM signal.

For the R(ight) command, we get the required reduction speed in the `value` variable. Then, we proportionally scale the 0-100 to the range 0-`currentSpeed`. In fact, the speed reduction of the right motor cannot be larger than the actual motor speed. In other words, when the command value is 255 (the maximum rotation speed), the speed of the right motor is reduced to 0, and the rover turns right. The same thing happens for the L(eft) command.

iOS code

In this chapter, we are going to look at the iOS application that remotely controls the rover. This application allows us to manually control the rover by the means of two sliders, that emulate the steering wheel and the throttle.

However, we are going to push the application a lot further by using the iOS device accelerometer and even voice commands. These techniques can be successfully applied to many other projects too.

Let's take a step at a time so that you can understand every important detail. We will start from the manual control. As always, the full code of this project can be downloaded from `https://www.packtpub.com/books/content/support`.

To have a better understanding of the explanations in the next paragraphs, open the downloaded code while reading.

Creating the Xcode project

We will create a new project as we already did in the previous chapters.
The following are the parameters for the new project:

- **Project Type**: Tabbed application
- **Product Name**: Rover
- **Language**: Objective-C
- **Devices**: Universal

In the project's options, we need to deselect the following options:

- **Landscape Right**
- **Landscape Left**

We do this because we are going to use the iOS accelerometer, and we don't want the screen to rotate when we rotate the device (see the following screenshot). To access the project's options, perform the following steps:

1. Select the project in the left pane of Xcode.

2. Select the **General** tab in the right pane.

We need additional code with the graphics components to show the rover inclination by the means of two gauges. The library can be downloaded from `https://github.com/sabymike/MSSimpleGauge`.

To install the additional code, perform the following steps:

1. Open the preceding link and click on the **Download ZIP** button on the right.
2. Unzip the downloaded ZIP file.
3. Open the `Gauges` folder and copy the files to the `Rover` folder of your project.

4. Select the **Rover** group in Xcode and right-click on it. Select **Add Files to "Rover"...**. Then, select the files that you just copied and click on **Add**. Make sure that **Copy items if needed** is selected (see the following screenshot):

5. In Xcode, select the files that you just added, right-click on them, select **New Group from Selection**, and then enter `Gauges`. This helps us keep the code organized.

6. To avoid a compilation error, open the file `MSArcLayer.h` and add `#import <UIKit/UIKit.h>` just before `#import <QuartzCore/QuartzCore.h>`.

We also need the library to accept voice commands (OpenEars), which can be downloaded from `http://www.politepix.com/openears/`.

Click on the **Download OpenEars** button. To install the library, you have to perform the following steps:

1. Uncompress the downloaded file.
2. Inside your downloaded distribution, there is a folder called `Framework`. Drag the `Framework` folder into your app project in Xcode.

Now that we have configured the required additional libraries, we can start creating the application.

The structure of this project is very close to the Pet Door Locker. So, we can reuse at least a part of the user interface and the code, by performing the following steps:

1. Select **FirstViewController.h** and **FirstViewController.m**, right-click on them, and click on **Delete** (see the following screenshot). Then, click on **Move to Trash**:

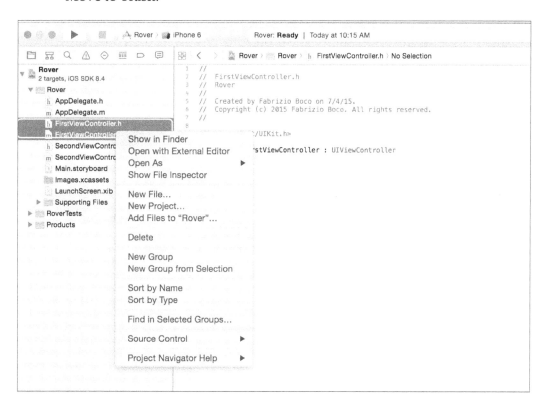

2. By using the same procedure, delete **SecondViewController** and **Main.storyboard**.

3. Open the **PetDoorLocker** project in Xcode.

4. Select the following files and drag and drop them to the **Rover** project:

    ◦ **PetDoorLockerViewController.h**

    ◦ **PetDoorLockerViewController.m**

    ◦ **BLEConnectionViewController.h**

    ◦ **BLEConnectionViewController.m**

    ◦ **Main.storyboard**

    Be sure that **Copy items if needed** is selected, and then click on **Finish**:

 If you have added icons to the Tab Bar, don't forget to drag and drop them too.

1. Rename `PetDoorLockerViewController` to `RoverViewController` by using the same procedure that we used in the previous chapters.

2. Open the **Main.storyboard** and locate the main **View** controller.

3. Delete the following GUI components:

    ◦ The **Door Status**, **Temperature**, **Label**, and **Lock** labels

    ◦ The **switch** component

4. Move the status view close to the **connect** button and update its Layout Constraints.

5. Add the GUI components and the related Layout Constraints, as shown in the following screenshot:

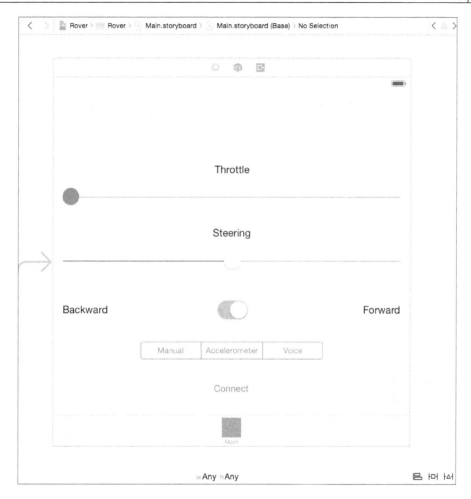

At the top, there are two UIViews for which, the size is 64 x 128 and the class is MSRangeGauge (change it in the Property Inspector)

5. For the **Throttle** slider, go to the Property Inspector and set the following values:

   ° **Minimum Value**: 0

   ° **Maximum Value**: 100

   ° **Current Value**: 0

   ° **Thumb Tin**: Red or whatever you like

6. For the **Steering** slider, go to the Property Inspector and set the following values:

    ◦ **Minimum Value**: 0

    ◦ **Maximum Value**: 200

    ◦ **Current Value**: 100

7. For the Segment Control, go to the Property Inspector and set **Segments** to 3.

8. Change the titles by double-clicking on each segment and entering the following values:

    ◦ Manual

    ◦ Accelerometer

    ◦ Voice

9. Select the View Controller container and in the Identity Inspector, change **Class** to **RoverViewController**.

10. In the `RoverViewController.h` file, add `#import  "MSRangeGauge.h"`.

11. Link the GUI components to the RoverViewController code, as follows:

```
@property (strong, nonatomic) IBOutlet UIView
*connectionStatus;
@property (strong, nonatomic) IBOutlet MSRangeGauge
*verticalIndicator;
@property (strong, nonatomic) IBOutlet MSRangeGauge
*horizontalIndicator;
@property (strong, nonatomic) IBOutlet UISlider
*throttleSlider;
@property (strong, nonatomic) IBOutlet UISlider
*steeringWheelSlider;
@property (strong, nonatomic) IBOutlet UISwitch
*directionSwitch;
@property (strong, nonatomic) IBOutlet UISegmentedControl
*modeSegment;
```

In your `RoverViewController.m`, there still are some references to the older project. Don't bother about these references. We will remove them in the following sections.

12. Link the Throttle slider to the method, as follows: `(IBAction)` `throttleChanged:(UISlider *)sender`.

13. Link the Steering slider to the method, as follows: `(IBAction)` `steeringWheelChanged:(UISlider *)sender`.

14. Link the forward/backward switch to the method, as follows: `(IBAction)` `directionChanged:(UISwitch *)sender`.

15. Link the mode segment controller with the method, as follows: `(IBAction)` `modeChange:(UISegmentedControl *)sender`.

Writing code for BLEConnectionViewController

Since we have copied this View Controller from the PetDoorLocker project, we don't need to change it.

We saved some work!

Writing code for RoverViewController

First, we have to remove the unnecessary code from the previous PetDoorLocker project that we don't need anymore, by performing the following steps:

1. Open `RoverViewController.m`.

2. Remove the following lines:
   ```
   @property (strong, nonatomic) IBOutlet UIView        *doorStatus;
   @property (strong, nonatomic) IBOutlet UILabel       *temperature;
   @property (strong, nonatomic) IBOutlet UISwitch
   *manualLockSwitch;
   ```

3. Remove the lines that refer to `_temperature` and `_doorStatus` in the code. Please refer to the downloaded code in case you have any doubt.

4. Empty the `dataReceived` function; we will rewrite it later:
   ```
   -(void)dataReceived:(NSString *)content {

   }
   ```

5. Completely remove the `switchChanged` function.

We can now start writing new code to control our rover.

Let's start from the easy part—receiving acceleration data from the rover. This data gives us information about the rover's inclination along its longitudinal and transverse axis.

This information will be shown by the two gauges (RMRangeGauges) that we added in the main screen. The gauges have to be initialized in the viewDidLoad method, as follows:

```
- (void)viewDidLoad {

    [super viewDidLoad];

    _centralManager = [[CBCentralManager alloc] initWithDelegate:self
queue:nil];

    _verticalIndicator.transform = CGAffineTransformMakeRotation(M_
PI/2);

    _verticalIndicator.minValue = 0;
    _verticalIndicator.maxValue = 200;
    _verticalIndicator.upperRangeValue = 130;
    _verticalIndicator.lowerRangeValue = 70;
    _verticalIndicator.value = 100;
    _verticalIndicator.fillArcFillColor = [UIColor colorWithRed:.9
green:.1 blue:.1 alpha:1];
    _verticalIndicator.rangeFillColor   = [UIColor colorWithRed:.2
green:.9 blue:.2 alpha:1];

    _horizontalIndicator.minValue = 0;
    _horizontalIndicator.maxValue = 200;
    _horizontalIndicator.upperRangeValue = 130;
    _horizontalIndicator.lowerRangeValue = 70;
    _horizontalIndicator.value = 100;
    _horizontalIndicator.fillArcFillColor = [UIColor colorWithRed:.9
green:.1 blue:.1 alpha:1];
    _horizontalIndicator.rangeFillColor   = [UIColor colorWithRed:.2
green:.9 blue:.2 alpha:1];

    ...

}
```

The code is very easy to understand, thanks to the self-documenting names of the methods.

Since the gauges are horizontal and don't have any features to show the needle vertically, we use the following instruction:

```
_verticalIndicator.transform = CGAffineTransformMakeRotation(M_PI/2);
```

With the preceding code, we rotate the first gauge by 90 degrees in order to have a better indication of the vertical inclination.

The inclination data is received by the dataReceived method as a string: <vertical inclination>:<horizontal inclination>, and the values are set to the two gauges, as follows:

```
- (void)dataReceived:(NSString *)content {

    NSArray *components = [content componentsSeparatedByString:@":"];
    if (components.count != 2) {
        return;
    }

    float x = [components[0] floatValue];
    float y = [components[1] floatValue];

    _verticalIndicator.value = 100+20*y;
    _horizontalIndicator.value = 100+20*x;
}
```

In the didDisconnectPeripheral method, which is called when the Bluetooth device disconnects, we have to reset the position of the two gauges by adding the following lines:

```
        _verticalIndicator.value = 100;
        _horizontalIndicator.value = 100;
```

Since we have three modes to operate the Rover — manually, by using the iOS accelerometer, and through voice commands — we are going to divide the code writing into three different sections to have a better understanding of the code.

Code to control the rover manually

For this scenario, we have to write the code that manages the Throttle slider to control the speed of the rover, the Steering slider to control its direction, and the switch to control its forward or backward movements.

The code for the Throttle slider is quite simple, since we only have to send a message to the rover in the form of T=<speed>, as follows:

```
- (IBAction)throttleChanged:(UISlider *)sender {

    NSInteger throttle = sender.value;

    NSString *msg = [NSString stringWithFormat:@"T=%ld",(long)
throttle];

    NSData* data;
    data=[msg dataUsingEncoding:NSUTF8StringEncoding];

    [_arduinoDevice writeValue:data forCharacteristic:_
sendCharacteristic type:CBCharacteristicWriteWithoutResponse];
}
```

The method that is used to control the direction is not that complex. It sends two messages depending on the position of the slider with respect to its middle—R=<speed> to turn the rover towards the right and L=<speed> to turn the rover towards the left:

```
- (IBAction)steeringWheelChanged:(UISlider *)sender {

    NSInteger steering = sender.value-100;

    NSString *msg;

    if (steering>0) {

        msg = [NSString stringWithFormat:@"R=%ld",(long)steering];
    }
    else {
        msg = [NSString stringWithFormat:@"L=%ld",(long)-steering];
    }

    NSData* data;
```

```
    data=[msg dataUsingEncoding:NSUTF8StringEncoding];

    [_arduinoDevice writeValue:data forCharacteristic:_
sendCharacteristic type:CBCharacteristicWriteWithoutResponse];
}
```

To complete the management of the two sliders, we need to change the
didConnectPeripheral method so that when the iOS device connects to
the Rover, the two sliders are reset to their initial position, as follows:

```
- (void)centralManager:(CBCentralManager *)central didConnectPeriphera
l:(CBPeripheral *)peripheral {

    _steeringWheelSlider.value = 100;
    _throttleSlider.value = 0;


    [peripheral discoverServices:@[[CBUUID UUIDWithString:NRF8001BB_
SERVICE_UUID]]];
}
```

The last method that we need to write is to control the forward/backward
direction. We need to send two simple messages to the rover— F is for forward
and B is for backward:

```
- (IBAction)directionChanged:(UISwitch *)sender {

    NSData* data;

    if (sender.on) {
        data=[@"F" dataUsingEncoding:NSUTF8StringEncoding];
    }
    else {
        data=[@"B" dataUsingEncoding:NSUTF8StringEncoding];
    }

    [_arduinoDevice writeValue:data forCharacteristic:_
sendCharacteristic type:CBCharacteristicWriteWithoutResponse];

    _throttleSlider.value = 0;
    [self throttleChanged:_throttleSlider];
}
```

To avoid undesirable behavior, each time we switch the direction, the rover is stopped, by setting the speed to 0 via the `[self throttleChanged: _throttleSlider]` method call.

We are now ready for the rover's first test.

Testing the Rover with manual driving

To perform the first rover test, you can use the following procedure:

1. Upload the Arduino code and check the console for any error message. If everything goes fine, Arduino is ready to take control over your rover.

2. Power up both the motors and Arduino itself.

3. Upload the iOS application to your device.

4. Go to the second tab to scan for the Bluetooth BLE breakout board.

5. Go to the first tab and progressively increase the Throttle slider. You should see the rover moving forward.

6. You can make it move right and left by moving the Steering slider.

7. Turn the direction switch, the rover stops. On increasing the speed again, the rover moves in the opposite direction it was moving before.

8. When going on a slope, you will see the needle of the vertical and/or the horizontal gauge move up or down and/or left or right.

9. When moving against an obstacle, the rover should stop before colliding against it.

Weird rover movements

If the rover moves in a wrong direction, you probably have not wired either or both of the motors properly. Swap the wires of the motor that is spinning in the wrong direction.

Weird gauge indications

If the gauges don't seem to be moving accordingly with the slope, don't worry too much about this for now. We are going to calibrate them later. For now, we only have to check whether the data is being properly transferred from the rover to the iOS device.

Missing gauge indications

If you cannot see any indication from the gauges, there may be an error in cabling the ADXL345 device. First double-check whether the Arduino console has any error message. If this doesn't help, remove the comments from the following two lines in the Arduino code (the main loop):

```
Serial.print("X: ")
Serial.print(event.acceleration.x); Serial.print("   ");
Serial.print("Y: "); Serial.print(event.
acceleration.y);
Serial.print("   ");Serial.println("m/s^2 ");
```

Connect the iOS device again and check the Arduino console. If you can see the printed numbers, then the accelerometer is working and you have to double-check the iOS code.

The rover hits obstacles

The distance sensor is sensitive to the shape, reflectance, and position of the obstacles and sometimes, it's not able to avoid them. This is the reason why different types of sensors are used at the same time, and in different positions in real-world robots. To check whether the sensor is properly connected and it's working as expected, you can remove the comment in the following line (in the main loop):

```
Serial.print("D: "); Serial.println(distance);
```

You should now be able to see the distance in the Arduino console.

Code for controlling the rover by the means of the iOS accelerometer

We are now going to improve our app by using the iOS accelerometer to control the steering and the throttle.

As we learned earlier, any deviation in the iOS device position can be measured and used to send appropriate commands to the rover with the help of the accelerometer that we mounted on the rover.

To access the accelerometer information, we need to use the `CMMotionManager` class. First, we add `#import <CoreMotion/CoreMotion.h>` to `RoverViewController.h`. Then, we create a property, as follows:

```
@interface RoverViewController ()

...

@property (strong, nonatomic) CMMotionManager        *motionManager;

...

@end
```

Finally, we initialize it in the `viewDidLoad` method, as follows:

```
- (void)viewDidLoad {

    [super viewDidLoad];

...

    _motionManager = [[CMMotionManager alloc] init];
...
}
```

The accelerometer is activated when the second button is selected on the segment controller, and the related method is called:

```
- (IBAction)modeChange:(UISegmentedControl *)sender {

    _throttleSlider.value = 0;
    _steeringWheelSlider.value = 100;

    if (_modeSegment.selectedSegmentIndex==1) {

        ...
        [self useAccelerometer];
    }

    ...
}
```

The `useAccelerometer` method actually activates the accelerometer:

```
-(void)useAccelerometer {

    [_motionManager setDeviceMotionUpdateInterval:0.2];
    [_motionManager startDeviceMotionUpdatesUsingReferenceFrame:
CMAttitudeReferenceFrameXArbitraryZVertical

toQueue:[NSOperationQueue mainQueue]

withHandler:^(CMDeviceMotion *motion, NSError *error) {

                                                 CMQuaternion
quat = motion.attitude.quaternion;

                                            [self sendAcce
lerometersCommands:quat];
                                        }];
}
```

The `[_motionManager setDeviceMotionUpdateInterval:0.2]` method instructs the motion manager to update our code with acceleration values every 0.2 seconds.

The next method actually starts updating our code of accelerometer values, which are received in the handler block.

Fortunately, iOS provide us with not only the actual value of acceleration along the three axes of the device, but also the quaternions. Don't be afraid of the name! They simply represent orientations and rotations of the iOS device in three dimensions (if you like mathematics, visit `https://en.wikipedia.org/wiki/Quaternions_and_spatial_rotation`). From them, you can easily calculate the two angles of the pitch and roll of the iOS device (see the following image).

If you have a liking for mathematics, visit `https://en.wikipedia.org/wiki/Conversion_between_quaternions_and_Euler_angles`:

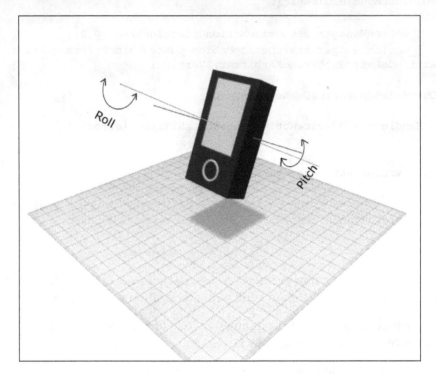

The roll is used to change the rover's direction of movement, and the pitch is used to control the throttle. The `- (void) sendAccelerometersCommands: (CMQuaternion) quad` method calculates the two angles and generates the messages that need to be sent to the rover, as we did in `throttleChanged` and `steeringWheelChanged`.

Driving the rover by the means of the iOS device movement

To test the rover, tap on the accelerometer, switch to forward, and hold the iOS device vertically. By moving the device forward around the pitch axis (see the previous image), the rover should start moving forward. The more the device moves forward, the more the rover's speed increases. By moving the device backward, the rover's speed reduces, and you can stop the rover by using this movement.

To turn the rover left or right, the device has to be turned left or right around the roll axis.

To make the commands more or less responsive, you can try to change the frequency at which the code is updated with new acceleration values, as follows:

```
[_motionManager setDeviceMotionUpdateInterval:0.2];
```

Don't forget that increasing the update interval increases the battery consumption. The tradeoff is strictly related to your rover and your needs.

Code for controlling the rover by voice commands

Voice recognition has been a challenge for years, but nowadays, you can add this feature to your application in a matter of minutes.

We have already added the required library. So, we can start adding the required code by performing the following steps:

1. Open `RoverViewController.h` and add `#import <OpenEars/OEEventsObserver.h>`.

2. Change the interface from `@interface RoverViewController : UIViewController <CBCentralManagerDelegate, CBPeripheralDelegate>` to `@interface RoverViewController : UIViewController <CBCentralManagerDelegate, CBPeripheralDelegate, OEEventsObserverDelegate>`.

3. Open `RoverViewController.m` and add the following imports:
```
#import <OpenEars/OELanguageModelGenerator.h>
#import <OpenEars/OEPocketsphinxController.h>
#import <OpenEars/OEAcousticModel.h>
#import <Slt/Slt.h>
#import <OpenEars/OEFliteController.h>
```

4. Add the following properties:
```
@property (strong, nonatomic) NSString          *lmPath;
@property (strong, nonatomic) NSString          *dicPath;

@property (strong, nonatomic) OEEventsObserver
*openEarsEventsObserver;
@property (strong, nonatomic) OEFliteController
*fliteController;
@property (strong, nonatomic) Slt               *slt;
```

5. Initialize the properties by adding the following code to the `viewDidLoad` method. The `words` array contains the voice commands that will be recognized. The rest of the code is from the documentation of the libraries:

```
NSMutableArray *words = [[NSMutableArray alloc]
initWithArray:@[@"RIGHT", @"LEFT", @"CENTER", @"FORWARD",
@"SLOWFORWARD", @"FASTFORWARD", @"BACKWARD", @"SLOWBACKWARD",
@"FASTBACKWARD",@"STOP"]];

    _fliteController = [[OEFliteController alloc] init];
    _slt = [[Slt alloc] init];

    OELanguageModelGenerator *lmGenerator =
[[OELanguageModelGenerator alloc] init];

    NSError *err=nil;
    NSString  *name = @"RoverVoiceControl";

    err = [lmGenerator generateLanguageModelFromArray:words
withFilesNamed:name forAcousticModelAtPath:[OEAcousticModel pathTo
Model:@"AcousticModelEnglish"]];

    if(err == nil) {

        _lmPath = [lmGenerator pathToSuccessfullyGeneratedLanguage
ModelWithRequestedName:name];
        _dicPath = [lmGenerator pathToSuccessfullyGeneratedDiction
aryWithRequestedName:name];

    } else {
        NSLog(@"Error: %@",[err localizedDescription]);
    }
```

6. Change the `modeChange` method in the following way (turn off the voice commands recognition and/or accelerometer when they are not being used):

```
- (IBAction)modeChange:(UISegmentedControl *)sender {

    _throttleSlider.value = 0;
    _steeringWheelSlider.value = 100;

    if (_modeSegment.selectedSegmentIndex==0) {
        [[OEPocketsphinxController sharedInstance] stopListening];
        [_motionManager stopDeviceMotionUpdates];
    }
```

```
    if (_modeSegment.selectedSegmentIndex==1) {

        [[OEPocketsphinxController sharedInstance] stopListening];
        [self useAccelerometer];
    }

    if (_modeSegment.selectedSegmentIndex==2) {

        [_motionManager stopDeviceMotionUpdates];
        [self useVoice];
    }
}
```

7. Add the following useVoice method. It activates the listening of voice commands and configures the library in order to call the pocketsphinxDidReceiveHypothesis delegate method when a voice command is recognized:

```
-(void)useVoice {

    [[OEPocketsphinxController sharedInstance] setActive:TRUE
error:nil];

    [[OEPocketsphinxController sharedInstance]
startListeningWithLanguageModelAtPath:_lmPath

dictionaryAtPath:_dicPath
                                                              ac
ousticModelAtPath:[OEAcousticModel pathToModel:@"AcousticModelEngl
ish"]

languageModelIsJSGF:NO];

    [[OEPocketsphinxController sharedInstance]
setSecondsOfSilenceToDetect:.7];
    [[OEPocketsphinxController sharedInstance]
setVadThreshold:3.0];

    _openEarsEventsObserver = [[OEEventsObserver alloc] init];
    [_openEarsEventsObserver setDelegate:self];
}
```

8. Add the `pocketsphinxDidReceiveHypothesis` method, which can be copied from the downloaded code. It doesn't do much other than formatting and sending commands to the rover much as we already did for the other modes. We just need to point out the following:

   ° The `hypothesis` parameter is a string with the recognized command

   ° The `[_fliteController say:hypothesis withVoice:self.slt]` call allows you to hear the recognized command that is pronounced by your iOS device

```objc
- (void) pocketsphinxDidReceiveHypothesis:(NSString *)
hypothesis recognitionScore:(NSString *)recognitionScore
utteranceID:(NSString *)utteranceID {

    [_fliteController say:hypothesis withVoice:self.slt];

    NSString *msg=nil;

    if ([hypothesis isEqualToString:@"FORWARD"]) {

        [_arduinoDevice writeValue:[@"F" dataUsingEncoding:
NSUTF8StringEncoding] forCharacteristic:_sendCharacteristic
type:CBCharacteristicWriteWithoutResponse];

        msg = [NSString stringWithFormat:@"T=%ld",60l];
    }

    ...

    NSData* data;
    data=[msg dataUsingEncoding:NSUTF8StringEncoding];

    [_arduinoDevice writeValue:data forCharacteristic:_
sendCharacteristic type:CBCharacteristicWriteWithoutRespon
se];
}
```

9. To complete the app, we have to add few lines in the `applicationDidEnterBackground` method in the `AppDelegate.m` file to disconnect from the rover when the app is sent to the background:

```
    UITabBarController *tabController = (UITabBarController *)_
window.rootViewController;
    RoverViewController *roverController = tabController.
viewControllers[0];

    [roverController disconnect];
```

Driving the rover by voice commands

To try this feature, you have to tap on Voice and then speak any of the available commands. When the application identifies a voice command, it pronounces the recognized command and the rover will start moving accordingly.

Please note that the voice recognition takes some time. Hence, the rover is not very responsive. The voice mode is more appropriate in wide space and for long navigation (whatever this could mean for a rover!)

If you experience low background noises that trigger speech recognition, you can raise the value in this call in the range of 1.5-3.5:

```
    [[OEPocketsphinxController sharedInstance] setVadThreshold:3.0]
```

To make the rover a little bit more responsive, you can try to reduce the time that the app should wait for after the speech ends to attempt recognizing the speech (the default value is 0.7 seconds), as follows:

```
    [[OEPocketsphinxController sharedInstance]
    setSecondsOfSilenceToDetect:.7];
```

Testing and tuning

We have already tested each mode of driving the rover, but we may still have unreliable readings from the accelerometer that is mounted on the rover itself.

To calibrate the accelerometer's readings, use the following procedure:

1. Make sure that the ADXL345 is firmly mounted on the rover, and its axes are parallel to the longitudinal and transverse axes of the rover.

2. Place the rover on a firm, flat surface and ensure that the rover is in plane with the help of a spirit level.

3. Comment out the following lines in the `loop` function of the Arduino code and upload it:

```
Serial.print("X: "); Serial.print(event.acceleration.x);
Serial.print(" ");
Serial.print("Y: "); Serial.print(event.acceleration.y);
Serial.print(" ");Serial.println("m/s^2 ");
```

4. Power Arduino via the USB cable and open the console.

5. Connect the iOS device. The acceleration reading appears on the console for both the axes. They should be 0 or very close to 0. If this is not the case, take 10 to 20 readings, calculate the average, and put these values in the `setup` function for the `xOffset` and `yOffset` variables.

Now, the readings from the accelerometer should be more consistent, and the two needles on the iOS devices should help you drive the rover on rough terrain, avoiding overturns.

How to go further

The following are some suggestions to improve the project:

1. Show the distance to the front obstacles in the iOS application.

2. Stop the rover when its inclination goes beyond a certain threshold.

3. Add more distance sensors or mount the distance sensor on a servo motor so that the rover can detect obstacles all around it.

4. Mount different types of sensors for better obstacle avoidance (for instance, ultrasound distance sensors or laser distance sensors).

5. Allow iOS device landscape orientation. Here's a hint—you need to get the actual iOS device orientation by using one of the orientation delegate methods (`willTransitionToTraitCollection`, `viewWillTransitionToSize`), which is provided by the `ViewController` class.

Voice recognition can be used on many projects because it's very simple to set up and works pretty well. You can start adding voice recognition to the Wi-Fi Power Plug project.

If you need a challenge, try to improve your rover by making it move autonomously, which can be done by making the rover become aware of its position. (A hint—you can use particle filters to accomplish this, but this is a very tough subject. The Google car is based on this and a lot of other things as well).

Summary

You were introduced to the fascinating world of robotics and controlled vehicles. You learned how to write Arduino code to control both the speed and the direction of rotation of DC motors, measure the distance with an analog infrared sensor, and measure acceleration along the three axes by using an accelerometer.

You learned how to use new graphical components such as the UISlider and the UISegmentedControl on an iOS device, and take advantage of the accelerometer that is sported by iOS devices. Moreover, you learned how to improve your projects with a very powerful and easy-to-use library for voice recognition and text to speech.

Don't forget that you now have a rover that you can drive without other people thinking that you are odd; you are not playing like you were a child, you are learning robotics! Have fun with your rover!

This chapter is quite long and the project was complex, but we can now catch our breath. In the next chapter, we will build a very simple but extremely powerful project. It controls the volume of your TV set by keeping it at almost the same level even if commercials are on air. Even if the project is pretty easy, you will learn a lot about IR transmitters and receivers, and digital signal processing.

5
TV Set Constant Volume Controller

I don't watch TV much, but when I do, I usually completely relax and fall asleep. I know that TV is not meant for putting you to sleep, but it does this to me. Unfortunately, commercials are transmitted at a very high volume and they wake me up. How can I relax if commercials wake me up every five minutes?

Can you believe it? During one of my naps between two commercials, I came up with a solution based on iOS and Arduino.

It's nothing complex. An iOS device listens to the TV set's audio, and when the audio level becomes higher than a preset threshold, the iOS device sends a message (via Bluetooth) to Arduino, which controls the TV set volume, emulating the traditional IR remote control. Exactly the same happens when the volume drops below another threshold. The final result is that the TV set volume is almost constant, independent of what is on the air. This helps me sleep longer!

The techniques that you are going to learn in this chapter are useful in many different ways. You can use an IR remote control for any purpose, or you can control many different devices, such as a CD/DVD player, a stereo set, Apple TV, a projector, and so on, directly from an Arduino and iOS device. As always, it is up to your imagination.

Constant Volume Controller requirements

Our aim is to design an Arduino-based device, which can make the TV set's volume almost constant by emulating the traditional remote controller, and an iOS application, which monitors the TV and decides when to decrease or increase the TV set's volume.

Hardware

Most TV sets can be controlled by an IR remote controller, which sends signals to control the volume, change the channel, and control all the other TV set functions.

IR remote controllers use a carrier signal (usually at 38 KHz) that is easy to isolate from noise and disturbances.

The carrier signal is turned on and off by following different rules (encoding) in order to transmit the 0 and 1 digital values.

The IR receiver removes the carrier signal (with a low pass filter) and decodes the remaining signal by returning a clear sequence of 0 and 1.

The IR remote control theory

You can find more information about the IR remote control at `http://bit.ly/1UjhsIY`.

Our circuit will emulate the IR remote controller by using an IR LED, which will send specific signals that can be interpreted by our TV set.

On the other hand, we can receive an IR signal with a phototransistor and decode it into an understandable sequence of numbers, by designing a demodulator and a decoder.

Nowadays, electronics is very simple; an IR receiver module (Vishay 4938) will manage the complexity of signal demodulation, noise cancellation, triggering, and decoding. It can be directly connected to Arduino, making everything very easy.

In the project in this chapter, we need an IR receiver to discover the coding rules that are used by our own IR remote controller (and the TV set).

Additional electronic components

In this project, we need the following additional components:

- IR LED Vishay TSAL6100
- IR Receiver module Vishay TSOP 4838
- Resistor 100Ω
- Resistor 680Ω
- Electrolytic capacitor 0.1μF

Electronic circuit

The following picture shows the electrical diagram of the circuit that we need for the project:

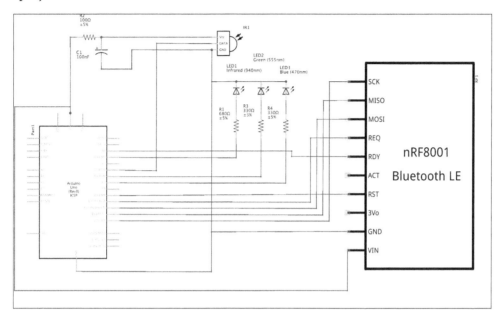

The IR receiver will be used only to capture the TV set's remote controller signals so that our circuit can emulate them.

However, an IR LED is constantly used to send commands to the TV set. The other two LEDs will show when Arduino increases or decreases the volume. They are optional and can be omitted.

As usual, the Bluetooth device is used to receive commands from the iOS device.

Powering the IR LED in the current limits of Arduino

From the datasheet of the TSAL6100, we know that the forward voltage is 1.35V. The voltage drop along R1 is then *5-1.35 = 3.65V*, and the current provided by Arduino to power the LED is about *3.65/680=5.3 mA*. The maximum current that is allowed for each PIN is 40 mA (the recommended value is 20 mA). So, we are within the limits. In case your TV set is far from the LED, you may need to reduce the R1 resistor in order to get more current (and the IR light). Use a new value of R1 in the previous calculations to check whether you are within the Arduino limits. For more information about the Arduino PIN current, check out `http://bit.ly/1JosGac`.

The following diagram shows how to mount the circuit on a breadboard:

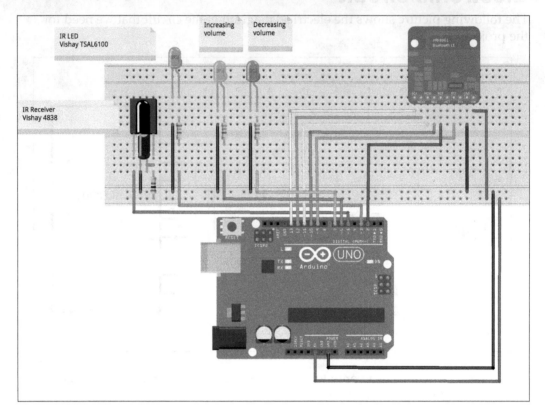

Arduino code

The entire code of this project can be downloaded from `https://www.packtpub.com/books/content/support`.

To understand better the explanations in the following paragraphs, open the downloaded code while reading them.

In this project, we are going to use the IR remote library, which helps us code and decode IR signals.

The library can be downloaded from `http://bit.ly/1Isd8Ay`, and installed by using the following procedure:

1. Navigate to the release page of `http://bit.ly/1Isd8Ay` in order to get the latest release and download the `IRremote.zip` file.

2. Unzip the file whatever you like.

3. Open the **Finder** and then the `Applications` folder (*Shift + Control + A*).

4. Locate the Arduino application.

5. Right-click on it and select **Show Package Contents**.

6. Locate the `Java` folder and then `libraries`.

7. Copy the `IRremote` folder (unzipped in step 2) into the `libraries` folder.

8. Restart Arduino if you have it running.

In this project, we need the following two Arduino programs:

- One is used to acquire the codes that your IR remote controller sends to increase and decrease the volume

- The other is the main program that Arduino has to run to automatically control the TV set volume

Let's start with the code that is used to acquire the IR remote controller codes.

Decoder setup code

In this section, we will be referring to the downloaded `Decode.ino` program that is used to discover the codes that are used by your remote controller.

Since the setup code is quite simple, it doesn't require a detailed explanation; it just initializes the library to receive and decode messages.

Decoder main program

In this section, we will be referring to the downloaded `Decode.ino` program; the main code receives signals from the TV remote controller and dumps the appropriate code, which will be included in the main program to emulate the remote controller itself.

Once the program is run, if you press any button on the remote controller, the console will show the following:

```
For IR Scope:
+4500 -4350 ...

For Arduino sketch:
unsigned int raw[68] = {4500,4350,600,1650,600,1600,600,1600,...};
```

The second row is what we need. Please refer to the *Testing and tuning* section for a detailed description of how to use this data.

Now, we will take a look at the main code that will be running on Arduino all the time.

Setup code

In this section, we will be referring to the `Arduino_VolumeController.ino` program. The setup function initializes the nRF8001 board and configures the pins for the optional monitoring LEDs.

Main program

The `loop` function just calls the `polACI` function to allow the correct management of incoming messages from the nRF8001 board.

The program accepts the following two messages from the iOS device (refer to the `rxCallback` function):

- `D` to decrease the volume
- `I` to increase the volume

The following two functions perform the actual increasing and decreasing of volume by sending the two `up` and `down` buffers through the IR LED:

```
void volumeUp() {
  irsend.sendRaw(up, VOLUME_UP_BUFFER_LEN, 38);
  delay(20);
}

void volumeDown() {
  irsend.sendRaw(down, VOLUME_DOWN_BUFFER_LEN, 38);
  delay(20);
  irsend.sendRaw(down, VOLUME_DOWN_BUFFER_LEN, 38);
  delay(20);
}
```

The up and down buffers, VOLUME_UP_BUFFER_LEN and VOLUME_DOWN_BUFFER_LEN, are prepared with the help of the Decode.ino program (see the *Testing and tuning* section).

iOS code

In this chapter, we are going to look at the iOS application that monitors the TV set volume and sends the volume down or volume up commands to the Arduino board in order to maintain the volume at the desired value.

The full code of this project can be downloaded from https://www.packtpub.com/books/content/support.

To understand better the explanations in the following paragraphs, open the downloaded code while reading them.

Creating the Xcode project

We will create a new project as we already did in the previous chapters. The following are the steps that you need to follow:

The following are the parameters for the new project:

- **Project Type**: Tabbed application
- **Product Name**: VolumeController
- **Language**: Objective-C
- **Devices**: Universal

To set a capability for this project, perform the following steps:

1. Select the project in the left pane of Xcode.
2. Select **Capabilities** in the right pane.

3. Turn on the **Background Modes** option and select **Audio and AirPlay** (refer to the following picture). This allows an iOS device to listen to audio signals too when the iOS device screen goes off, or the app goes in the background:

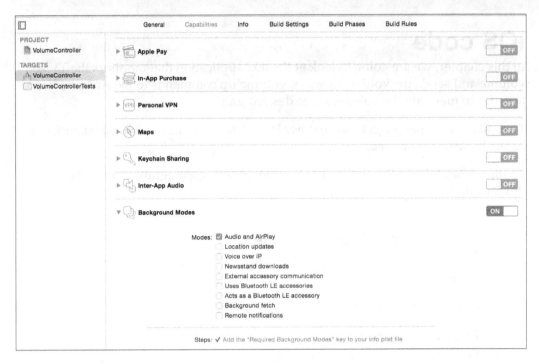

Since the structure of this project is very close to the Pet Door Locker, we can reuse a part of the user interface and the code by performing the following steps (for more details, go back to the iOS guider rover project in *Chapter 4, iOS Guided Rover*, where we did almost the same thing):

1. Select **FirstViewController.h** and **FirstViewController.m**, right-click on them, click on **Delete**, and select **Move to Trash**.

2. With the same procedure, delete **SecondViewController** and **Main.storyboard**.

3. Open the **PetDoorLocker** project in Xcode.

4. Select the following files and drag and drop them to this project (refer to the following picture).
   ○ **BLEConnectionViewController.h**
   ○ **BLEConnectionViewController.m**
   ○ **Main.storyboard**

   Ensure that **Copy items if needed** is selected and then click on **Finish**.

5. Copy the icon that was used for the BLEConnectionViewController view controller.

6. Create a new View Controller class and name it `VolumeControllerViewController`.

7. Open the **Main.storyboard** and locate the main View Controller.

8. Delete all the graphical components.

9. Open the Identity Inspector and change the **Class** to **VolumeControllerViewController**.

Now, we are ready to create what we need for the new application.

Designing the user interface for VolumeControllerViewController

This view controller is the main view controller of the application, and contains just the following components:

* The switch that turns on and off the volume control
* The slider that sets the desired volume of the TV set

Once you have added the components and their layout constraints, you will end up with something that looks like the following screenshot:

Once the GUI components are linked with the code of the view controller, we end with the following code:

```
@interface VolumeControllerViewController ()

@property (strong, nonatomic) IBOutlet UISlider    *volumeSlider;

@end
```

and with:
```
- (IBAction)switchChanged:(UISwitch *)sender {
…
}
- (IBAction)volumeChanged:(UISlider *)sender {
…
}
```

Writing code for BLEConnectionViewController

Since we copied this View Controller from the Pet Door Locker project, we don't need to change it apart from replacing the key, which was used to store the peripheral UUID, from `PetDoorLockerDevice` to `VolumeControllerDevice`.

We saved some work!

Now, we are ready to work on the VolumeControllerViewController, which is much more interesting.

Writing code for VolumeControllerViewController

This is the main part of the application; almost everything happens here.

We need some properties, as follows:

```
@interface VolumeControllerViewController ()

@property (strong, nonatomic) IBOutlet UISlider    *volumeSlider;

@property (strong, nonatomic) CBCentralManager    *centralManager;
@property (strong, nonatomic) CBPeripheral        *arduinoDevice;
```

```
@property (strong, nonatomic) CBCharacteristic    *sendCharacteristic;

@property (nonatomic,strong) AVAudioEngine         *audioEngine;

@property float                                    actualVolumeDb;
@property float                                    desiredVolumeDb;
@property float                                    desiredVolumeMinDb;
@property float                                    desiredVolumeMaxDb;

@property NSUInteger                               increaseVolumeDelay;

@end
```

Some are used to manage the Bluetooth communication and don't need much explanation. The `audioEngine` is the instance of `AVAudioEngine`, which allows us to transform the audio signal captured by the iOS device microphone in numeric samples. By analyzing these samples, we can obtain the power of the signal that is directly related to the TV set's volume (the higher the volume, the greater the signal power).

Analog-to-digital conversion

The operation of transforming an analog signal into a digital sequence of numbers, which represent the amplitude of the signal itself at different times, is called analog-to-digital conversion. Arduino analog inputs perform exactly the same operation. Together with the digital-to-analog conversion, it is a basic operation of digital signal processing and storing music in our devices and playing it with a reasonable quality. For more details, visit `http://bit.ly/1N1QyXp`.

The `actualVolumeDb` property stores the actual volume of the signal measured in dB (short for decibel).

Decibel (dB)

The decibel (dB) is a logarithmic unit that expresses the ratio between two values of a physical quantity. Referring to the power of a signal, its value in decibel is calculated with the following formula:

$$P_{dB} = 10 * Log_{10}\left(\frac{P}{P_0}\right)$$

Here, P is the power of the signal and P_0 is a reference power. You can find out more about decibel at `http://bit.ly/1LZQM0m`. We have to point out that if $P < P_0$, the value of P_{dB} if lower of zero. So, decibel values are usually negative values, and 0dB indicates the maximum power of the signal.

The `desiredVolumeDb` property stores the desired volume measured in dB, and the user controls this value through the volume slider in the main tab of the app; `desiredVolumeMinDb` and `desiredVolumeMaxDb` are derived from the `desiredVolumeDb`.

The most significant part of the code is in the `viewDidLoad` method (refer to the downloaded code).

First, we instantiate the `AudioEngine` and get the default input node, which is the microphone, as follows:

```
_audioEngine = [[AVAudioEngine alloc] init];
AVAudioInputNode *input = [_audioEngine inputNode];
```

The `AVAudioEngine` is a very powerful class, which allows digital audio signal processing. We are just going to scratch its capabilities.

AVAudioEngine

You can find out more about AVAudioEngine by visiting `http://apple.co/1kExe35` (AVAudioEngine in practice) and `http://apple.co/1WYG6Tp`.

The `AVAudioEngine` and other functions that we are going to use require that we add the following imports:

```
#import <AVFoundation/AVFoundation.h>
#import <Accelerate/Accelerate.h>
```

By installing an audio tap on the bus for our input node, we can get the numeric representation of the signal that the iOS device is listening to, as follows:

```
[input installTapOnBus:0 bufferSize:8192 format:[input
inputFormatForBus:0] block:^(AVAudioPCMBuffer* buffer, AVAudioTime*
when) {
...
...
}];
```

As soon as a new buffer of data is available, the code block is called and the data can be processed. Now, we can take a look at the code that transforms the audio data samples into actual commands to control the TV set:

```
for (UInt32 i = 0; i < buffer.audioBufferList->mNumberBuffers; i++) {

    Float32 *data = buffer.audioBufferList->mBuffers[i].mData;
    UInt32 numFrames = buffer.audioBufferList->mBuffers[i].
mDataByteSize / sizeof(Float32);

  // Squares all the data values
    vDSP_vsq(data, 1, data, 1, numFrames*buffer.audioBufferList-
>mNumberBuffers);

            // Mean value of the squared data values: power of the
  signal
    float meanVal = 0.0;
    vDSP_meanv(data, 1, &meanVal, numFrames*buffer.audioBufferList-
>mNumberBuffers);

    // Signal power in Decibel
    float meanValDb = 10 * log10(meanVal);

    _actualVolumeDb = _actualVolumeDb + 0.2*(meanValDb - _
actualVolumeDb);


    if (fabsf(_actualVolumeDb) < _desiredVolumeMinDb && _
centralManager.state == CBCentralManagerStatePoweredOn && _
sendCharacteristic != nil) {

        //printf("Decrease volume\n");

        NSData* data=[@"D" dataUsingEncoding:NSUTF8StringEncoding];
```

```
      [_arduinoDevice writeValue:data forCharacteristic:_
sendCharacteristic type:CBCharacteristicWriteWithoutResponse];

      _increaseVolumeDelay = 0;
}

    if (fabsf(_actualVolumeDb) > _desiredVolumeMaxDb && _
centralManager.state == CBCentralManagerStatePoweredOn && _
sendCharacteristic != nil) {

      _increaseVolumeDelay++;
}

    if (_increaseVolumeDelay > 10) {

      //printf("Increase volume\n");

      _increaseVolumeDelay = 0;

      NSData* data=[@"I" dataUsingEncoding:NSUTF8StringEncoding];
            [_arduinoDevice writeValue:data forCharacteristic:_
sendCharacteristic type:CBCharacteristicWriteWithoutResponse];
      }
}
```

In our case, the `for` cycle is executed just once, because we have just one buffer and we are using only one channel.

The power of a signal, represented by N samples, can be calculated by using the following formula:

$$P = \frac{1}{N} \sum_{n=0}^{N} v(n)$$

Here, v is the value of the nth signal sample.

Because the power calculation has to performed in real time, we are going to use the following functions, which are provided by the Accelerated Framework:

- `vDSP_vsq`: This function calculates the square of each input vector element
- `vDSP_meanv`: This function calculates the mean value of the input vector elements

The Accelerated Framework

The Accelerated Framework is an essential tool that is used for digital signal processing. It saves you time in implementing the most used algorithms and mostly providing implementation of algorithms that are optimized in terms of memory footprint and performance. More information on the Accelerated Framework can be found at `http://apple.co/1PYIKE8` and `http://apple.co/1JCJWYh`.

Eventually, the signal power is stored in `_actualVolumeDb`. When the modulus of `_actualVolumeDb` is lower than the `_desiredVolumeMinDb`, the TV set's volume is too high, and we need to send a message to Arduino to reduce it. Don't forget that `_actualVolumeDb` is a negative number; the modulus decreases this number when the TV set's volume increases. Conversely, when the TV set's volume decreases, the `_actualVolumeDb` modulus increases, and when it gets higher than `_desiredVolumeMaxDb`, we need to send a message to Arduino to increase the TV set's volume.

During pauses in dialogues, the power of the signal tends to decrease even if the volume of the speech is not changed. Without any adjustment, the increasing and decreasing messages are continuously sent to the TV set during dialogues. To avoid this misbehavior, we send the volume increase message only when the signal power stays over the threshold for some time (when `_increaseVolumeDelay` is greater than 10).

We can take a look at the other view controller methods that are not complex.

When the view belonging at the view controller appears, the following method is called:

```
- (void)viewDidAppear:(BOOL)animated {

    [super viewDidAppear:animated];

    NSError* error = nil;

    [self connect];

    _actualVolumeDb = 0;
    [_audioEngine startAndReturnError:&error];

    if (error) {
        NSLog(@"Error %@",[error description]);
    }

}
```

In this function, we connect to the Arduino board and start the audio engine in order to start listening to the TV set.

When the view disappears from the screen, the `viewDidDisappear` method is called, and we disconnect from the Arduino and stop the audio engine, as follows:

```
-(void)viewDidDisappear:(BOOL)animated {

    [self viewDidDisappear:animated];

    [self disconnect];

    [_audioEngine pause];
}
```

The method that is called when the switch is operated (`switchChanged`) is pretty simple:

```
- (IBAction)switchChanged:(UISwitch *)sender {

    NSError* error = nil;

    if (sender.on) {
        [_audioEngine startAndReturnError:&error];

        if (error) {
            NSLog(@"Error %@", [error description]);
        }
        _volumeSlider.enabled = YES;
    }
    else {
        [_audioEngine stop];
        _volumeSlider.enabled = NO;
    }
}
```

The method that is called when the volume slider changes is as follows:

```
- (IBAction)volumeChanged:(UISlider *)sender {

    _desiredVolumeDb = 50.*(1-sender.value);
    _desiredVolumeMaxDb = _desiredVolumeDb + 2;
    _desiredVolumeMinDb = _desiredVolumeDb - 3;
}
```

We just set the desired volume and the lower and upper thresholds.

The other methods that are used to manage the Bluetooth connection and data transfer don't require any explanation, because they are exactly like in the previous projects.

Testing and tuning

We are now ready to test our new amazing system and spend more and more time watching TV (or taking more and more naps!) Let's perform the following procedure:

1. Load the `Decoder.ino` sketch and open the Arduino IDE console.

2. Point your TV remote controller to the TSOP4838 receiver and press the button that increases the volume. You should see something like the following appearing on the console:

   ```
   For IR Scope:
   +4500 -4350 …
   ```

   ```
   For Arduino sketch:
   unsigned int raw[68] = {4500,4350,600,1650,600,1600,600,1600,…};
   ```

3. Copy all the values between the curly braces.

4. Open the `Arduino_VolumeController.ino` and paste the values for the following:

   ```
   unsigned int up[68] = {9000, 4450, …..,};
   ```

5. Check whether the length of the two vectors (68 in the example) is the same and modify it, if needed.

6. Point your TV remote controller to the TSOP4838 receiver, and press the button that decreases the volume. Copy the values and paste them for:

   ```
   unsigned int down[68] = {9000, 4400, …..,};
   ```

7. Check whether the length of the two vectors (68 in the example) is the same and modify it, if needed.

8. Upload the `Arduino_VolumeController.ino` to Arduino and point the IR LED towards the TV set.

9. Open the iOS application, scan for the nRF8001, and then go to the main tab.

10. Tap on connect and then set the desired volume by touching the slider.

11. Now, you should see the blue LED and the green LED flashing. The TV set's volume should stabilize to the desired value.

To check whether everything is properly working, increase the volume of the TV set by using the remote control; you should immediately see the blue LED flashing and the volume getting lower to the preset value. Similarly, by decreasing the volume with the remote control, you should see the green LED flashing and the TV set's volume increasing.

Take a nap, and the commercials will not wake you up!

How to go further

The following are some improvements that can be implemented in this project:

1. Changing channels and controlling other TV set functions.
2. Catching handclaps to turn on or off the TV set.
3. Adding a button to mute the TV set.
4. Muting the TV set on receiving a phone call.

Anyway, you can use the IR techniques that you have learned for many other purposes. For example, you can modify the rover project in *Chapter 4, iOS Guided Rover*, to control the robot via an IR remote controller. Take a look at the other functions provided by the IRremote library to learn the other provided options. You can find all the available functions in the IRremote.h that is stored in the IRremote library folder.

On the iOS side, try to experiment with the AV Audio Engine and the Accelerate Framework that is used to process signals.

Summary

This chapter focused on an easy but useful project and taught you how to use IR to transmit and receive data to and from Arduino. There are many different applications of the basic circuits and programs that you learned here.

On the iOS platform, you learned the very basics of capturing sounds from the device microphone and the DSP (digital signal processing). This allows you to leverage the processing capabilities of the iOS platform to expand your Arduino projects.

The next chapter will be amazing. You are going to open your garage door magically; you don't even need to touch your iOS device to do this. You will also learn a lot about the iBeacon technology. Your imagination will be your only limit!

6

Automatic Garage
Door Opener

This project is about an emerging technology called **iBeacon** that is based on the Bluetooth BLE communication protocol. Basically, an iBeacon is a small device that continuously transmits a unique coded signal. An iOS device can detect iBeacon to determine whether it is more or less near an iBeacon and trigger actions.

iBeacon is an Apple technology

iBeacon is a technology that was invented by Apple, and the protocol has not been disclosed (if you can find any information on the Internet, Google is your friend). So, to work with iBeacon devices, you need an iOS device and the API provided by Apple.

There is another standard that is emerging and which can also work with Android devices—AltBeacon (for more information, visit `http://bit.ly/1KsXD17`). The AltBeacon site has a lot of useful information for iOS too.

We are going to use these technologies to open our garage door as soon as we get close enough to it. You may wonder what the difference is between this and a traditional garage remote control or the numerous remote control apps that are available for iOS devices. The main difference is that everything happens automatically (automagically?); you don't even need to touch your phone. Believe it or not, you don't even need the controller app running.

Don't have a garage?

This project can be used to open any kind of door, but you may need to adapt or change your lock to something that can be controlled by an electric signal. Alternatively, you can use this project to control internal/external lights. There's no chance of getting bored!

Let's get started by having a closer look at iBeacon.

iBeacon – a technical overview

An iBeacon is a small device that, leveraging the Bluetooth BLE, establishes a region around itself. Any iOS device that supports Bluetooth BLE can determine whether it has entered or exited the region and roughly estimate the distance from the iBeacon.

We can put an iBeacon close to each museum artwork and write an iOS application that shows artwork information as soon as a visitor gets close to it. This is a typical example of how iBeacon is used.

iBeacon is univocally identified by three values—a UUID (a 16 bytes universal identifier), major (2 bytes), and minor (2 bytes), which are constantly transmitted over the Bluetooth signal.

All that you need to know about iBeacon

You can find all that you need to know at `https://developer.apple.com/ibeacon/` and `https://developer.apple.com/ibeacon/Getting-Started-with-iBeacon.pdf`.

Referring to the next diagram, you can see that there is a region around the iBeacon (the iBeacon region). The iOS application that registers to this region (with the UUID, major, and minor of the iBeacon) receives an "entering" notification when the iOS device crosses this region's border and enters it, and an "exiting" notification when the iOS device crosses the region's border and exits it.

We will discuss the details of this in the following sections:

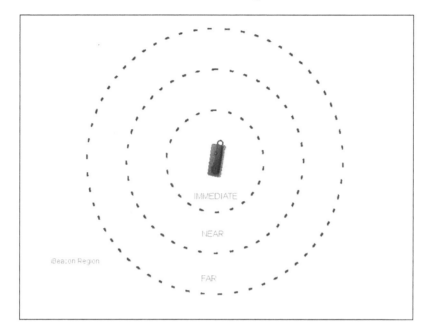

In open air, the iBeacon region is about 30 meters wide, but this size widely depends on the hardware design, configuration, obstacles, and the way of installation.

When inside the region, an iOS device can continuously monitor the distance from the iBeacon device. This distance is returned as Far, Near, or Immediate. Apple doesn't declare the actual size of these distances, most likely because they are really influenced by the iBeacon's transmitting power, obstacles between the iBeacon and the iOS device, the iOS device orientation, and other factors. Anyway, this information can be used to change the application's behavior with the distance from iBeacon.

iBeacon distance calculation

An iOS device can find out the signal strength of the iBeacon. It can calculate the distance from the iBeacon by using a formula that describes the signal attenuation with distance. Unfortunately, since an iBeacon signal has a lot of fluctuations and its propagation is influenced by a lot of factors, the formula gives a very poor estimation of the distance. Probabilistic techniques have to be used to get the estimation of the distance to use this value for practical applications. We cannot cover these techniques in this book.

An interesting feature of how the iBeacon notifications are handled by the iOS is that even if the application is not running, the entering and exiting notifications are received and they start the application. To save the iOS device power, the application is started just for a few seconds (for about 3 seconds) and then paused. Then, the application has to complete every operation in this short interval.

I have to point out that usually, an iOS app receives an entering notification almost as soon as it crosses the iBeacon region. Conversely, the exiting notification may be received even a few minutes after the region border is crossed.

The garage door opener requirements and design constraints

Knowing the capabilities of the iBeacon, it's not hard to imagine how a garage door opener may work. The iBeacon is installed just behind the garage door. An Arduino with the BLE board listens for commands from the iOS device and controls the garage door opener. An iOS app sends an "Open" command when entering into the iBeacon region and a "Close" command when exiting the iBeacon region. Very simple, isn't it?

Unfortunately, we have to face the potentially long delay when receiving an exiting notification. When driving a car, you can cover a great distance traveling away from your garage in a few minutes; an iOS device cannot connect to the Arduino in such a long time.

Moreover, after taking a look at the next picture, you may realize that the iBeacon region may also cover only a part of the house. We obviously don't want that walking around the house with the iOS device in our pockets, the garage door would open and close randomly.

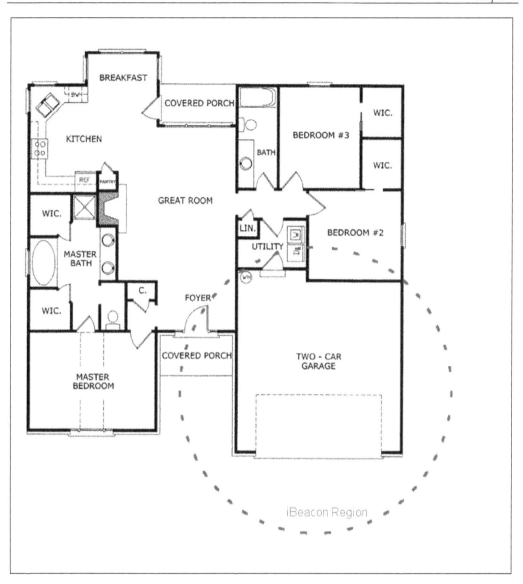

A simple solution could be adding as many iBeacons as needed across the house, all with the same UUID, major, and minor, in order to extend the iBeacon region.

However, this solution will be too expensive. We are going to use another power feature of the iOS to overcome these two issues—geofencing:

We can register a geofencing region (using the latitude and longitude of the center and the radius) that covers our entire house. On crossing the geofencing region's border, we get a notification (or even better, two notifications—one when entering the region and the other when exiting).

You may wonder why we don't simply use only a geofencing region to open the garage door. The reason behind not doing this is that the geofencing notifications are not very precise. They may have a very long delay and may not work in all areas. By combining the two technologies, we get exactly what we want to achieve.

We eventually overcame all the design challenges that involve tracking the entering of the device in the iBeacon region and its exiting from the geofencing region. This is how the automatic garage door opener works:

1. Let's assume that we are outside both the regions. When we cross the geofencing region, nothing happens, because the app recognizes only the exiting notification.

2. As soon as we cross the iBeacon region a few meters away from the garage, the iOS device gets a notification, and it sends an opening command to the Arduino board, which controls the door, thereby opening the door.

3. The door is automatically closed after a short time. This allows us to park the car inside the garage.

4. From now on, the action of crossing the iBeacon region is ignored. Then, in case we cross the iBeacon region while moving around the house, the garage will not be opened.

5. When we leave, we eventually cross the geofencing region. This event again enables the receiving of the iBeacon entering notification. We are in the same state that was described in step 1.

We can describe the behavior of the app by using the following state diagram (strictly speaking, it's a Mealy State Machine; for more information, visit `http://bit.ly/1hmZs3V`):

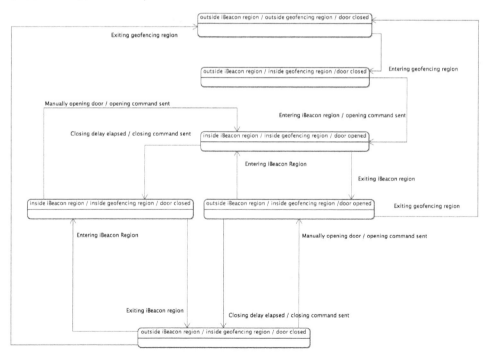

We obviously don't want anyone else to open our garage. So, each authorized user must identify themselves by using a PIN (personal identification number). Assigning a new PIN, you allow others to get access to your garage. You can revoke access just by removing the PIN from the authorization list.

Then, the app has to manage the PINs, and we need a master PIN, which is wired into the Arduino code. Only the person who knows the master PIN is authorized to manage PINs.

A security warning!

Even if the iOS app requires a PIN in order to open the garage door, the app is not completely secure since it doesn't provide any encryption mechanism. Anyone can access the PIN by using a Bluetooth protocol sniffer. It's not so easy, but it's possible. You have been warned! Making the communication secure could be a good chance to learn more about encryption and how to implement encryption using the little memory and little processing power available on Arduino.

Hardware

The main hardware component that we need is the iBeacon. We have used the one that is shown at `http://redbear.net`. Many products are available on the market at almost any price. When choosing an iBeacon, make sure that it is compatible with the iBeacon Apple protocol, since there are products that are not.

An iOS device as an iBeacon

If you have two iOS devices, you can use one of them to act as an iBeacon that running via one of the apps that are available on the iTunes store. I have published my own app for this purpose, which can be found at `http://apple.co/1hmZt80`.

Depending on the iBeacon enclosure, it can be mounted inside or outside the garage. It is better if it is placed at a higher position. Usually, the battery that powers the iBeacon should last for at least one year. Consider battery replacement anyway while positioning it.

Additional electronic components

In this project, we will need the following additional components:

- A BJT transistor: P2N2222 (see the details in the text)

- An NMOSFET transistor BS170 (see the details in the text)
- A 10K resistor (see the details in the text)
- A 1.5K resistor (see the details in the text)
- A diode 1N4001 (see the details in the text)
- A relay: coil voltage 5 and contact current max 1A (see the details in the text)

Electronic circuit

The following picture shows the electric diagram of the electronic circuit that we need for the project:

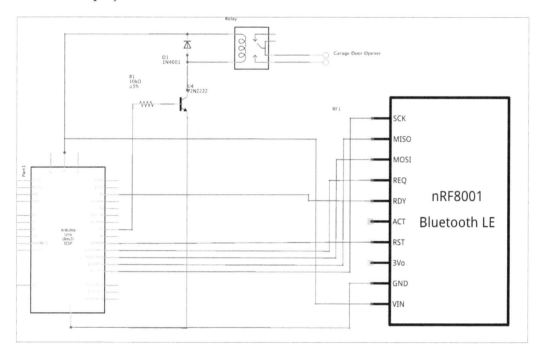

Usually, the relay current is about 40 mA, which exceeds the maximum amount of current that Arduino can provide. There are relays that draw less current, but to avoid burning the Arduino, we can use a transistor to power the relay. When the relay is turned off, the energy stored in the coil is discharged against the transistor as reverse current, and this can damage the transistor. The diode (a flyback diode) shorts this current, protecting the transistor from damage.

The following diagram shows how to mount the circuit on a breadboard:

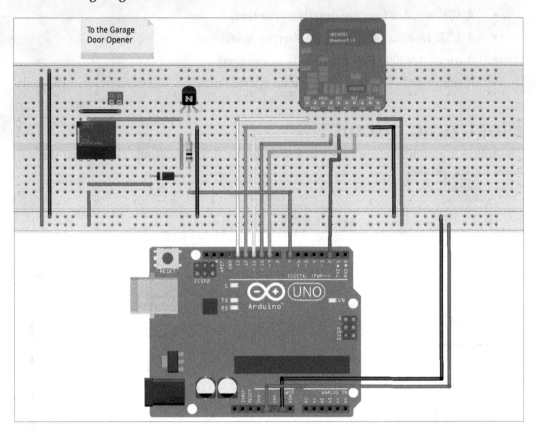

A relay is an electromechanical device that drains current, and it is subject to faults. So, you can use a more reliable circuit by using N-MOSFET. This alternative circuit is shown in the following diagram:

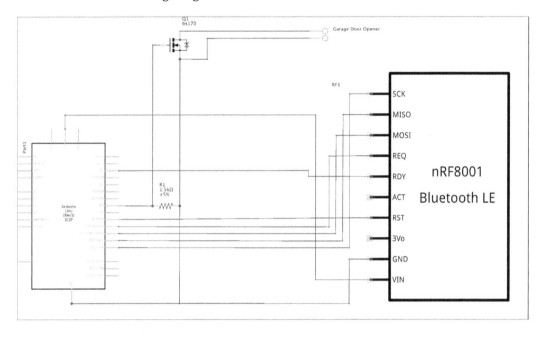

The following diagram shows how to mount the circuit on a breadboard:

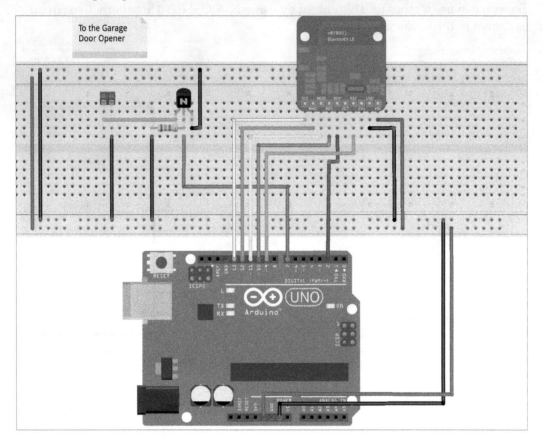

The R1 resistor pulls down the gate of the MOSFET. So, it turns off when the driving pin (7) is LOW and the pin is floating like that during the powering on phase of Arduino.

You may need to adjust the value of R1 in the range of 1K to 1M in order to ensure that the MOSFET doesn't turn on when not expected to.

RFduino instead of Arduino

For this project, we can use RFduino instead of Arduino (for more information, visit `http://www.rfduino.com`). It is compatible with Arduino, and also includes the Bluetooth device and the related software stack. Moreover, it can act as an iBeacon at the same time. One device could cover all the hardware requirements for the project. It doesn't have the EEPROM that we are going to use to store PINs, but we can store them in flash. The original version of this project was on RFduino and was a commercial product. For this book, I chose to work with Arduino to avoid buying another piece of hardware and because RFduino requires a more complex code to use Bluetooth for receiving commands and acting as an iBeacon at the same time. You can try to build the project on RFduino yourself. This could be a good opportunity to learn more. Check out RFduino. It's an amazing product!

Arduino code

The full code of this project can be downloaded from `https://www.packtpub.com/books/content/support`.

For a better understanding of the explanations in the next paragraphs, open the downloaded code while reading them.

In this project, we are going to use EEPROM to store PINS. In fact, this memory doesn't lose its content when it's not powered.

To store a PIN, we use the first character to indicate whether it is used or not and the last five characters to store the actual PIN (which is exactly five characters long). PINs are stored sequentially in EEPROM, starting from address 0.

Setup code

Please refer to the downloaded code. Since the setup code is quite simple, it doesn't require a detailed explanation.

The setup code is not much different from the setup code of other projects. Let's take a look at the EEPROM initialization instead:

```
void setup() {



    // EEPROM INITIALIZATION - FIRST TIME ONLY

    for (int i = 0; i < 6*NUMBER_OF_PINS; i++)
      EEPROM[i] = 0;

    // Set the master PIN

    EEPROM[0] = 1;
    EEPROM[1] = '1';
    EEPROM[2] = '2';
    EEPROM[3] = '3';
    EEPROM[4] = '4';
    EEPROM[5] = '5';


    ...


}
```

The `for` loop initializes the EEPROM locations that are used to initialize PINs to 0. This loop has to be executed only the first time the code is executed, otherwise, it clears the stored PINS. In the *Testing and tuning* section, we will provide more details on this.

The last few lines write the master PIN from location 1 (the master PIN is 12345 in the example). The location 0 is set to 1 to indicate that the next five locations are used to store a PIN.

Main program

The loop function is very easy. It only checks whether the door has been opened, and after this, CLOSING_DOOR_INTERVAL closes it by calling the pulseOutput function that pulses the relay for 300 ms.

The rest of the Arduino code is used to react to the messages that are received in the rxCallback function.

Each message is made up of a PIN (five characters) followed by a few other characters. The PIN is checked, and if it's not recognized, the message is rejected.

The main message is used to open the garage door: <PIN>O=1, where <PIN> is the 5 characters long PIN assigned to the user who is opening the door. When it's received, it pulses the relay and opens the door.

All the other messages, which are as follows, are related to PIN management:

- P, when the iOS device requests the list of the current PINs stored in the Arduino EEPROM.

- A, when the iOS device needs to add a new PIN. The next 5 bytes after the command are the actual PIN.

- E, when the iOS device needs to update an existing PIN. The first byte after the command is the index of the PIN that needs to be edited, which is followed by 5 bytes of the new PIN.

- D, when the iOS device needs to delete an existing PIN. The first byte after the command is the index of the PIN that needs to be deleted.

The functions, which implement each command, don't require much explanation. Note that the master PIN (PIN at the location 0 in EEPROM) is never transferred to the iOS application. To change the master PIN, it has to be changed directly in the code. The printPins function that dumps all the stored PINs can help you understand how the functions work. Comment out the calls that are already in the code.

iOS code

In this chapter, we are going to look at the iOS application that monitors the iBeacon region and the geofencing region and sends a command to open the door. The same application manages the PINs that can be assigned to relatives, guests, and friends in order to open your garage.

The application can also be used to open the garage door manually like a traditional remote control.

The full code of this project can be downloaded from `https://www.packtpub.com/books/content/support`.

To understand better the explanations in the next paragraphs, open the downloaded code while reading.

Creating the Xcode project

We will create a new project as we have done in the previous chapters. The following are the steps that you need to perform:

The following are the parameters for the new project:

- **Project Type**: Tabbed application
- **Product Name**: GarageiBeacon
- **Language**: Objective-C
- **Devices**: Universal

We have to set a capability for this project, as follows:

1. Select the project in the left pane of Xcode.
2. Select **Capabilities** in the right pane.

3. Turn on the **Background Modes** option and select **Location updates** (see the following screenshot):

Once more, since the structure of this project is very close to the Pet Door Locker, we can reuse a part of the user interface and code by following these steps (for more details, go back to the iOS guided rover project in *Chapter 4, iOS Guided Rover*, where we did almost the same thing as this):

1. Select **FirstViewController.h** and **FirstViewController.m**, right-click on them, click on **Delete**, and select **Move to Trash**.

2. Using the same procedure, delete **SecondViewController** and **Main.storyboard**.

3. Open the **PetDoorLocker** project in Xcode.

4. Select the following files and drag and drop them to this project:

    ° **BLEConnectionViewController.h**

    ° **BLEConnectionViewController.m**

    ° **Main.storyboard**

    Ensure that **Copy items if needed** is selected and then click on **Finish**.

5. Copy the icon that was used for the BLEConnectionViewController view controller.

6. Create a new View Controller class called `GarageViewController`.

7. Open the **Main.storyboard** and locate the main View Controller.

8. Delete all the graphical components.

9. Open the **Identity Inspector** and change the **Class** to **GarageViewController**.

10. In order to make location properly ask for authorization, we need to add a new file. Do this by navigating to **File | New | File...** and then select **iOS - Resource** and **Strings File**. Click on **Next** and enter the name of the file, `InfoPlist`. Finally, click on **Create**.

11. Open the newly created file and enter the following line:

```
NSLocationAlwaysUsageDescription = "This is required in order to
make Garage iBeacon working properly.";
```

Now, we are ready to create the new application!

Designing the user interface for BLEConnectionViewController

We have to add many components to this view controller to add our personal PIN and information related to the geofencing region.

By now, you should be a master in adding UIKit components and the related Layout Constraints. So, we won't spend much time on this subject. Your final result should look similar to the one in the following picture. Anyway, you can always refer to the downloaded code, in case you need to:

We also need to set the delegate outlet of the **PIN** text field to the BLEConnectionViewController in order to know when it changes. You can do this by using the Connection Inspector.

For the PIN text field, we need to mask the values that will be entered. To do this, perform the following steps:

1. Select the field.
2. Open the Attributes Inspector.
3. Select the **Secure Text Entry** checkbox.

Linking the new components to the code, you should end up with the following:

```
@interface BLEConnectionViewController ()

@property (strong, nonatomic) IBOutlet UILabel        *deviceUUIDLabel;

@property (strong, nonatomic) IBOutlet UITextField *pinField;
@property (strong, nonatomic) IBOutlet UITextField *longitudeField;
@property (strong, nonatomic) IBOutlet UITextField *latitudeField;
@property (strong, nonatomic) IBOutlet UISegmentedControl
*houseRegionSizeSegment;

@property (strong, nonatomic) CBCentralManager        *centralManager;
@property (strong, nonatomic) NSTimer                 *scanningTimer;

@end
```

The **Set House Position** button will be used to set the center of the geofencing region around the house, and the **House Region Size** segment will be used to set the radius of the geofencing region itself.

The button is connected to the method, as follows:

```
- (IBAction)startLocating:(id)sender {

}
```

The segment is connected to the following:

```
- (IBAction)regionSizeChanged:(UISegmentedControl *)sender {

}
```

Designing the user interface for GarageViewController

This view controller is the main view controller of the application, and it should contain the button to manually open/close the garage door, just in case!

Since this is a learning project, we added some components to give the user a lot more information about their position with respect to the iBeacon and geofencing regions.

The GUI should look like the following screenshot:

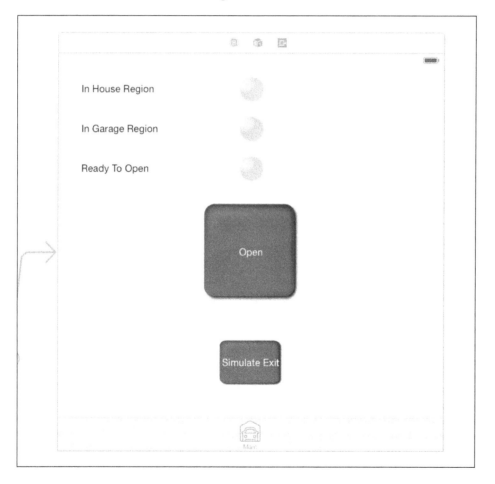

This time, the two buttons have a background (you can copy it from the downloaded code; it is named `buttonBackground.png`). To add it, select the button, open the **Attribute Inspector**, and select **buttonBackground.png** for the **Background** (see the following screenshot):

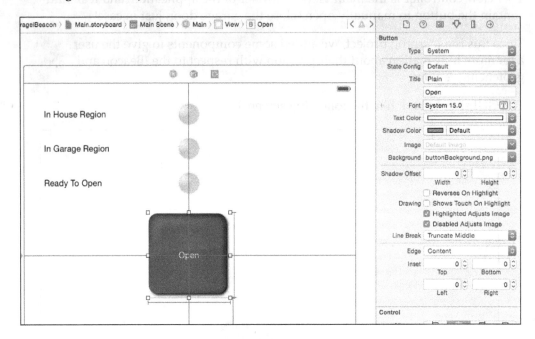

Don't forget to change the **Text Color** to white.

You can also copy the images for the three LEDs (`blueLED.png` and `grayLED.png`) from the downloaded project.

Once you have linked the GUI components to the code, you should end up with the following:

```
@interface GarageViewController ()

@property (strong, nonatomic) IBOutlet UIImageView
*houseRegionIndicator;
@property (strong, nonatomic) IBOutlet UIImageView
*garageRegionIndicator;
@property (strong, nonatomic) IBOutlet UIImageView
*readyToOpenIndicator;

@end
```

Moreover, the two buttons are linked to the two methods respectively, as follows:

```
- (IBAction)manualOperation:(UIButton *)sender {
}
```

and:

```
- (IBAction)simulateHomeRegionExit:(UIButton *)sender {
}
```

Designing the user interface for PinsViewController

We need another view controller to manage the PINs. Create and link it to the main view controller like we did in the previous projects and embed it into a Navigation Controller. To do this, select the new view controller and navigate to **Editor | Embed In Navigation Controller**. This creates a Navigation Bar, where we can drop a button for adding the PINs (see the circled area in the following screenshot).

The GUI components are shown in the following screenshot. Basically, they are a Table View to show the enabled PINs and a field where one can enter the Master PIN. Only the person who knows the Master PIN can manage the other pins:

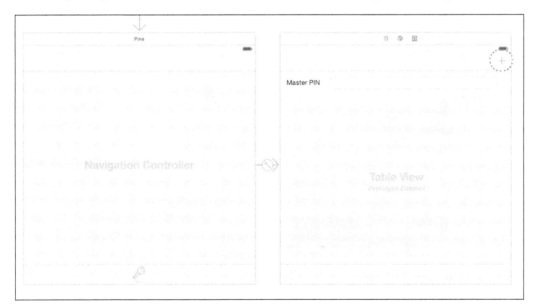

Before moving on to the next section, perform the following steps:

1. Create a new class named PinsViewController that is inherited from UIViewController.

2. Select the PinsViewController in the storyboard, open the Identity Inspector, and select PinsViewController for **Class**.

3. Open the Connections Inspector (browse **View | Utilities | Show Connections Inspector**).

4. Select the Table View, drag the dataSource, and delegate outlets to the PinsViewController class (see the following screenshot). This tells the Table View to ask for items to show and inform about the events in the PinsViewController:

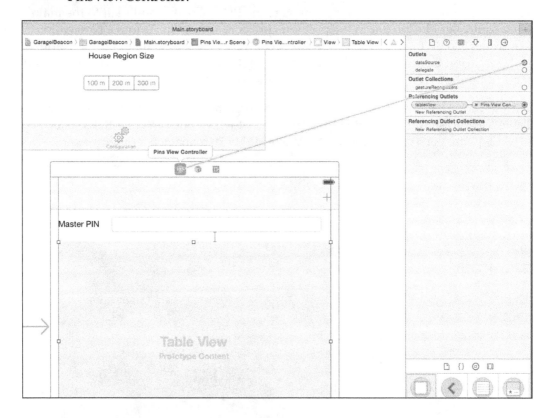

5. Set the delegate outlet of the master PIN field to the view controller.

6. Set the **Secure Text Entry** checkbox for the master PIN.

7. Linking the GUI components with the code, you should end up with the following:

```
@interface PinsViewController ()

@property (strong, nonatomic) IBOutlet UITableView *tableView;
@property (strong, nonatomic) IBOutlet UITextField *pinField;

@end
```

and with the method:

```
- (IBAction)addPin:(id)sender {
}
```

Writing code for BLEConnectionViewController

Since we copied this View Controller from the Pet Door Locker project, we need to make only a few changes.

First, we need to open the **BLEConnectionViewController.h** file and add the following import:

```
#import <CoreLocation/CoreLocation.h>
```

We also need to make a change to the following lines:

```
@interface BLEConnectionViewController : UIViewController <
CBCentralManagerDelegate>


@end
```

Change the preceding lines to the following:

```
@interface BLEConnectionViewController : UIViewController
<CLLocationManagerDelegate, CBCentralManagerDelegate>


@end
```

Then, open the `BLEConnectionViewController.m` to make the rest of the changes. Let's add a new property, as follows:

```
@property (strong, nonatomic) CLLocationManager     *locationManager;
```

The location manager allows us to get the geographical coordinates (through the GPS receiver of our iOS device) of our house. This will be used to create the geofencing region.

To initialize the location manager, we have to change the `viewDidAppear` method to the following:

```
- (void)viewDidAppear:(BOOL)animated {

    [super viewDidAppear:animated];

    _locationManager = [[CLLocationManager alloc] init];
    [_locationManager requestAlwaysAuthorization];

    _centralManager = [[CBCentralManager alloc] initWithDelegate:self
    queue:nil];
}
```

The location manager has to be authorized by the user to work. For this reason, we need to issue the following method call:

```
[_locationManager requestAlwaysAuthorization];
```

Calling this method, the iOS starts an authorization request to the user that contains the message we added to the InfoPlist file we created in the previous section. The code that is needed to start receiving the position of your own house is as follows:

```
- (IBAction)startLocating:(id)sender {

    _locationManager.delegate = self;
    _locationManager.distanceFilter = kCLDistanceFilterNone;
    _locationManager.desiredAccuracy = kCLLocationAccuracyBest;
    [_locationManager startUpdatingLocation];
}
```

The code doesn't require any explanation.

Once the GPS receiver has located the position, the following method is called, and we can store the longitude and latitude of the house:

```
- (void)locationManager:(CLLocationManager *)manager
didUpdateLocations:(NSArray *)locations {

    [manager stopUpdatingLocation];
    CLLocation *currentLocation = [locations objectAtIndex:0];

    _latitudeField.text = [NSString stringWithFormat:@"%f",currentLoca
tion.coordinate.latitude];
    _longitudeField.text = [NSString stringWithFormat:@"%f",currentLoc
ation.coordinate.longitude];

    NSUserDefaults *userDefaults = [NSUserDefaults
standardUserDefaults];

    [userDefaults setObject:[NSNumber numberWithFloat:currentLocation.
coordinate.latitude] forKey:@"HouseLatitude"];
    [userDefaults setObject:[NSNumber numberWithFloat:currentLocation.
coordinate.longitude] forKey:@"HouseLongitude"];
    [userDefaults synchronize];

    sleep(2);   // To be sure that monitoring of region started - To
avoid kCLErrorDomain error 5
}
```

Please note that once the coordinates are available, we stop the location manager to update them (`[manager stopUpdatingLocation]`). We do this to save batteries and because we do not check the coordinates anymore to know whether we are inside or outside the region around the house. See the following sections to understand how we get this information.

When we change the size of the geofencing region, the following method is called:

```
- (IBAction)regionSizeChanged:(UISegmentedControl *)sender {

    NSUserDefaults *userDefaults = [NSUserDefaults
standardUserDefaults];

    [userDefaults setObject:[NSNumber numberWithFloat:sender.
selectedSegmentIndex+1] forKey:@"HouseSize"];
    [userDefaults synchronize];
}
```

This is where we store the size of the region itself.

The last method that we need to write is used to store the personal PIN once it is modified:

```objc
- (BOOL)textFieldShouldReturn:(UITextField *)textField {

    [textField resignFirstResponder];

    NSUserDefaults *userDefaults = [NSUserDefaults
standardUserDefaults];

    [userDefaults setObject:textField.text
forKey:@"GarageiBeaconPIN"];
    [userDefaults synchronize];

    return YES;
}
```

To initialize the values of the text fields when the view controller is started, change the `viewDidLoad` method to the following:

```objc
- (void)viewDidLoad {

    [super viewDidLoad];

    _deviceUUIDLabel.text = [[NSUserDefaults standardUserDefaults] obj
ectForKey:@"GarageiBeaconDevice"];
    _pinField.text = [[NSUserDefaults standardUserDefaults] objectForK
ey:@"GarageiBeaconPIN"];

    _latitudeField.text = [[[NSUserDefaults standardUserDefaults]
objectForKey:@"HouseLatitude"] stringValue];
    _longitudeField.text = [[[NSUserDefaults standardUserDefaults]
objectForKey:@"HouseLongitude"] stringValue];
}
```

When the view controller is not shown on the screen, we can free both the location manager and the central manager, as follows:

```objc
- (void)viewDidDisappear:(BOOL)animated {
    [super viewDidDisappear:animated];

    _centralManager = nil;
    _locationManager = nil;
}
```

Now, we are ready to work on the GarageViewController, which is much more interesting.

Writing code for GarageViewController

Since we use both the location manager and the central manager, we need to update **GarageViewController.h** with the required includes and protocols ending with:

```
#import <CoreLocation/CoreLocation.h>
#import <CoreBluetooth/CoreBluetooth.h>


@interface GarageViewController : UIViewController
<CLLocationManagerDelegate, CBCentralManagerDelegate,
CBPeripheralDelegate>


@end
```

Then, we open the **GarageViewController.m** to add the code to manage the notifications from the geofencing regions and the iBeacon region, and to send an opening message to the Arduino.

Add the following properties:

```
@property (nonatomic,strong) CLLocationManager      *locationManager;
@property (nonatomic,strong) CBCentralManager       *centralManager;

@property (strong, nonatomic) CBPeripheral          *arduinoDevice;
@property (strong, nonatomic) CBCharacteristic
*sendCharacteristic;

@property                     BOOL                  insideHouse;
```

Then, we can add the code used to manage the Bluetooth communication with Arduino. This code is almost the same as what we used in the previous chapters. So, we don't have to spend much time on it. We just point out the following:

- Since we don't receive any data from Arduino via Bluetooth, the didUpdateValueForCharacteristic function can be removed, and we don't need to look for the characteristic that was used to receive data

- Once the application gets connected to Arduino, it immediately sends the opening command

The didDiscoverCharacteristicsForService method is slightly different from what we used in the other projects:

```objc
- (void)peripheral:(CBPeripheral *)peripheral didDiscoverCharacteristicsForService:(CBService *)service error:(NSError *)error {

    if (error) {

        NSLog(@"Error %@",[error localizedDescription]);

        return;
    }

    for (CBService *service in peripheral.services) {

        if ([service.UUID.UUIDString isEqualToString:NRF8001BB_SERVICE_UUID]) {

            for (CBCharacteristic *characteristic in service.characteristics) {

                if ([characteristic.UUID.UUIDString isEqualToString:NRF8001BB_CHAR_TX_UUID]) {

                    _sendCharacteristic = characteristic;

                    // Device connected - Sending opening command

                    NSData      *data;
                    NSString    *msg;

                    msg = [[NSString alloc] initWithFormat:@"%@ O=1",[[NSUserDefaults standardUserDefaults] objectForKey:@"GarageiBeaconPIN"]];
                    data=[msg dataUsingEncoding:NSUTF8StringEncoding];

                    [_arduinoDevice writeValue:data forCharacteristic:_sendCharacteristic type:CBCharacteristicWriteWithoutResponse];

                    // Disconnects

                    [_centralManager cancelPeripheralConnection:_arduinoDevice];
```

```
                }
            }
        }
    }
}
```

Once the view controller is started, we have to initialize the location manager and create the geofencing region and iBeacon region, if they are not yet created:

```
- (void)viewDidLoad {

    [super viewDidLoad];

    _centralManager = [[CBCentralManager alloc] initWithDelegate:self
queue:nil];

    _locationManager = [[CLLocationManager alloc] init];
    _locationManager.delegate = self;

    _locationManager.desiredAccuracy = kCLLocationAccuracyBest;
    _locationManager.distanceFilter = kCLDistanceFilterNone;
    _locationManager.activityType = CLActivityTypeOther;

    [_locationManager requestAlwaysAuthorization];

    for (CLRegion *region in _locationManager.monitoredRegions) {

        [_locationManager requestStateForRegion:region];
    }

    _arduinoDevice = nil;

    // Monitoring change of UserDefaults

    [[NSUserDefaults standardUserDefaults] addObserver:self

forKeyPath:@"HouseLongitude"
                                               options:NSKeyValueObse
rvingOptionNew
                                               context:NULL];

    [[NSUserDefaults standardUserDefaults] addObserver:self
```

```
                                         forKeyPath:@"HouseSize"
                                         options:NSKeyValueObse

    rvingOptionNew

                                         context:NULL];


        [self addObserver:self
            forKeyPath:@"insideHouse"
                options:NSKeyValueObservingOptionNew
                context:NULL];
    }
```

Once the location manager is initialized and authorized by the user, the following method is called and the iBeacon region is created (we will discuss the actual iBeacon region creation later in this section):

```
- (void)locationManager:(CLLocationManager *)manager didChangeAuthoriz
ationStatus:(CLAuthorizationStatus)status {

    if (status == kCLAuthorizationStatusAuthorizedAlways && _
centralManager.state == CBCentralManagerStatePoweredOn) {
        [self createGaregeRegionIfNeeded];
    }
}
```

Then we call a method:

```
[[NSUserDefaults standardUserDefaults] addObserver:self

forKeyPath:@"HouseLongitude"

                                         options:NSKeyValueObse
rvingOptionNew

                                         context:NULL];
```

This method activates a key-value observer in user defaults, where information about the geofencing region is stored.

Key-value observing

For more information about key-value observing, take a look at the Apple documentation at http://apple.co/1PZ6aJm.

Now, every time the `HouseLongitude` changes, the `observeValueForKeyPath` method is called, and this allows us to create or update the geofencing region. Note that `HouseLongitude` changes in the BLEConnectionViewController when the user locates the house. The KVO technique allows you to automatically keep the geofencing region updated. The KVO is also set for the `HouseSize` user default property so that the region gets updated when the user changes the size of the region itself.

We can also set the KVO for the `insideHouse` property. We will talk about this later on.

Let's take a look at the `observeValueForKeyPath` method where the geofencing region is actually created:

```
-(void)observeValueForKeyPath:(NSString *)aKeyPath ofObject:(id)
anObject change:(NSDictionary *)aChange context:(void *)aContext {


    if ([aKeyPath isEqualToString:@"insideHouse"]) {

        if (_insideHouse)
            _readyToOpenIndicator.image = [UIImage
imageNamed:@"grayLED.png"];
        else
            _readyToOpenIndicator.image = [UIImage
imageNamed:@"blueLED.png"];

        return;
    }


    CLLocationCoordinate2D center;

    center.latitude = [[[NSUserDefaults standardUserDefaults]
objectForKey:@"HouseLatitude"] floatValue];
    center.longitude = [[[NSUserDefaults standardUserDefaults]
objectForKey:@"HouseLongitude"] floatValue];

    double radius = [[[NSUserDefaults standardUserDefaults]
objectForKey:@"HouseSize"] doubleValue];
    radius = (radius == 0) ? 1 : radius;
```

```
        //NSLog(@"Latitude %f Longitude %f Radius %f",center.latitude,
    center.longitude, radius);

        CLCircularRegion *houseRegion = [[CLCircularRegion alloc]
    initWithCenter:center
    radius:100. * radius
    identifier:@"House Region"];
        houseRegion.notifyOnEntry = YES;
        houseRegion.notifyOnExit = YES;

        [_locationManager startMonitoringForRegion:houseRegion];
        [_locationManager requestStateForRegion:houseRegion];
    }
```

To create the geofencing region, we need to define its center, which is the location of the house, and a radius that is manually chosen. Then, call the following:

```
    [_locationManager startMonitoringForRegion:houseRegion];
```

The iOS knows that we need to receive a notification as soon as we enter or exit the geofencing region. Therefore, we call:

```
    [_locationManager requestStateForRegion:houseRegion];
```

We request the iOS to establish whether we are inside or outside the geofencing region and call the `locationManager:(CLLocationManager *)manager didDetermineState:(CLRegionState)state forRegion:(CLRegion *)region` method to inform us about the same (we'll talk about this method later).

Let's see how iBeacon is created instead. Remember that when the location manager is authorized and the central manager, which manages the Bluetooth connections and communication, is turned on, the following method is called in order to create the iBeacon region:

```
    -(void)createGaregeRegionIfNeeded {

        NSArray *regions = [_locationManager.monitoredRegions allObjects];

        NSPredicate *p = [NSPredicate predicateWithFormat:@"identifier ==
    %@",@"Garage Region"];
        NSArray *garageRegions = [regions filteredArrayUsingPredicate:p];
```

```
    if (garageRegions.count == 0 && _centralManager.state ==
CBCentralManagerStatePoweredOn) {

        NSUUID *beaconUUID = [[NSUUID alloc]
initWithUUIDString:@"00000000-0000-0000-0000-0000000000FF"];

        CLBeaconRegion *beaconRegion = [[CLBeaconRegion alloc] initWit
hProximityUUID:beaconUUID

major:0

minor:1

identifier:@"Garage Region"];
        beaconRegion.notifyEntryStateOnDisplay = YES;

        [_locationManager startMonitoringForRegion:beaconRegion];
        [_locationManager requestStateForRegion:beaconRegion];
    }
}
```

The creation of the region is quite similar to that of the geofencing region. This time, the region has an UUID, a major, and a minor instead of a center and a radius.

Once the iBeacon region is created, we ask the iOS to start monitoring it (`[_locationManager startMonitoringForRegion:beaconRegion]`) and immediately tell us whether we are inside or outside the region itself (`[_locationManager requestStateForRegion:beaconRegion]`).

Now, let's take a look at the most important part of the code – the code that actually manages the regions' boundary crossing and sends the command of opening the garage door.

Every time we enter a region, the `didEnterRegion` method is called:

```
- (void)locationManager:(CLLocationManager *)manager
didEnterRegion:(CLRegion *)region  {


    UILocalNotification* localNotification = [[UILocalNotification
alloc] init];
    localNotification.fireDate = nil;
    localNotification.alertBody = [NSString
stringWithFormat:@"Entering %@",region.identifier];
```

```
        localNotification.timeZone = [NSTimeZone defaultTimeZone];
        localNotification.soundName = @"Chime.aiff";
        [[UIApplication sharedApplication] presentLocalNotificationNow:loc
    alNotification];

        if ([region.identifier isEqualToString:@"Garage Region"]) {

            [_locationManager startRangingBeaconsInRegion:(CLBeaconRegion
    *)region];

            if (_insideHouse)
                return;

            NSString *deviceIdentifier = [[NSUserDefaults
    standardUserDefaults] objectForKey:@"GarageiBeaconDevice"];

            if (deviceIdentifier!=nil && _arduinoDevice==nil) {

                NSArray *devices = [_centralManager
    retrievePeripheralsWithIdentifiers:@[[CBUUID UUIDWithString:deviceIde
    ntifier]]];
                if (devices.count == 0) {
                    return;
                }

                _arduinoDevice = devices[0];
                _arduinoDevice.delegate = self;
            }

            [_centralManager connectPeripheral:_arduinoDevice
    options:nil];

            [self setInsideHouse:YES];
        }
    }
```

The first few lines send a local notification to the user to inform them that the region border has been crossed going into it.

If you are entering the iBeacon region and you are inside the house (_insideHouse = YES), nothing happens. This means that if the iBeacon region cannot cover the entire house and if you exit the iBeacon region by moving inside your house, you don't open the garage door unexpectedly.

If you are not in the house, the app gets connected to Arduino (via Bluetooth) and the garage door opens. Don't forget that the actual opening command is sent into the `didDiscoverCharacteristicsForService` method.

Forget `[_locationManager startRangingBeaconsInRegion:(CLBeaconRegion *)region]` for now.

Enabling local notifications

In order to send local notifications, they have to be authorized by the user. To do this, we need to call the `[application registerUs erNotificationSettings:[UIUserNotificationSettings settingsForTypes:UIUserNotificationTypeAlert|UIUse rNotificationTypeBadge|UIUserNotificationTypeSound categories:nil]]` method as soon as the application starts. The `didFinishLaunchingWithOptions` method is the place where we call it.

Every time we get into a region, the `didExitRegion` method is called:

```
- (void)locationManager:(CLLocationManager *)manager
didExitRegion:(CLRegion *)region {

UILocalNotification* localNotification = [[UILocalNotification alloc]
init];
    localNotification.fireDate = nil;
    localNotification.alertBody = [NSString stringWithFormat:@"Exiting
%@",region.identifier];
    localNotification.timeZone = [NSTimeZone defaultTimeZone];
    localNotification.soundName = @"Chime.aiff";
    [[UIApplication sharedApplication] presentLocalNotificationNow:loc
alNotification];

    [_locationManager stopRangingBeaconsInRegion:(CLBeaconRegion *)
region];

    if ([region.identifier isEqualToString:@"House Region"]) {

        [self setInsideHouse:NO];
    }
}
```

After sending a local notification, if we are exiting the geofencing region, we can set the `insideHouse` property to `NO` so that when we enter the iBeacon region again, the opening command is sent. Forget `[_locationManager stopRangingBeaconsInReg ion:(CLBeaconRegion *)region]` for now.

Why don't we use the traditional code (`_insideHouse = YES`) to set the property? In the `viewDidLoad` method, we set an observer for the property so that every time it changes, the `observeValueForKeyPath` is called. The traditional code doesn't start the `observeValueForKeyPath` method, and we need to use `[self setInsideHouse:NO]` instead. When the `observeValueForKeyPath` is called because `insideHouse` changes, we update the image of the `readyToOpen`, keeping the user informed about whether the app will send the opening command on entering the iBeacon region or not. To do this, we need these few lines of code into the `observeValueForKeyPath` method:

```
if ([aKeyPath isEqualToString:@"insideHouse"]) {

    if (_insideHouse)
        _readyToOpenIndicator.image = [UIImage
imageNamed:@"grayLED.png"];
    else
        _readyToOpenIndicator.image = [UIImage
imageNamed:@"blueLED.png"];

    return;
}
```

Working in the background

What makes the iBeacon technology along with geofencing particularly interesting is that `didEnterRegion` and `didExitRegion` are also called when the application is either running in the background, or not even running. Unfortunately, when the application is not running, the iOS starts it and keeps it running for some period of time (for about 3s) to save the batteries. So, any kind of action that is required to respond to the event has to be very quick. In the code, we just connect to the Arduino and send a few bytes to it, and this takes up much less time than 3s.

The last relevant method that we need to write is `didDetermineState`, which is called to find out where the device is with respect to a region (calling `[_locationManager startMonitoringForRegion:beaconRegion]` or `[_locationManager startMonitoringForRegion:houseRegion]`) or when the iOS recognizes that something has changed. In this function, we update the indicators that visually inform the user that they are in one of the monitored regions, as follows:

```
- (void)locationManager:(CLLocationManager *)manager didDetermineState
:(CLRegionState)state forRegion:(CLRegion *)region {

    switch (state) {

        case CLRegionStateInside:
            NSLog(@"Inside %@",region.identifier);
            break;

        case CLRegionStateOutside:
            NSLog(@"Outside %@",region.identifier);
            break;

        case CLRegionStateUnknown:
            NSLog(@"Unknown %@",region.identifier);
            break;
    }

    if ([region.identifier isEqualToString:@"Garage Region"]) {

        if (state==CLRegionStateInside) {

            _garageRegionIndicator.image = [UIImage
imageNamed:@"blueLED.png"];
        }
        else {

            _garageRegionIndicator.image = [UIImage
imageNamed:@"grayLED.png"];
        }
    }

    if ([region.identifier isEqualToString:@"House Region"]) {

        if (state==CLRegionStateInside) {
```

```
            _houseRegionIndicator.image = [UIImage
    imageNamed:@"blueLED.png"];
        }
        else {

            [self setInsideHouse:NO];
            _houseRegionIndicator.image = [UIImage
    imageNamed:@"grayLED.png"];
        }
    }


    }
```

Note that the visual information (inside the iBeacon and geofencing region, ready to open the garage door) is not strictly needed. We put that in the app to make you experiment with iBeacon and geofencing.

The last two methods are as follows:

- `manualOperation`: This manually opens the garage door by sending the opening command.
- `simulateHomeRegionExit`: This simulates the exit from the geofencing region by manually setting the `insideHouse` property to NO. This can be useful in the debugging phase or if you wish to learn how the application works without having to actually drive away from your house (which we did too many times!).

The code is very simple and doesn't require much explanation:

```
- (IBAction)manualOperation:(UIButton *)sender {

    NSString *deviceIdentifier = [[NSUserDefaults
standardUserDefaults] objectForKey:@"GarageiBeaconDevice"];

    if (deviceIdentifier!=nil && _arduinoDevice==nil) {

        NSArray *devices = [_centralManager
retrievePeripheralsWithIdentifiers:@[[CBUUID UUIDWithString:deviceIde
ntifier]]];
        if (devices.count == 0) {
            return;
        }
        _arduinoDevice = devices[0];
        _arduinoDevice.delegate = self;
    }

    if (_arduinoDevice != nil) {
```

```
        [_centralManager connectPeripheral:_arduinoDevice
options:nil];
    }
}

- (IBAction)simulateHomeRegionExit:(UIButton *)sender {

    //_insideHouse = NO; // This doesn't fire the KVO !
    [self setInsideHouse:NO];
}
```

The very last two methods (we promise!) that you need to look at are as follows:

- didRangeBeacons: This method is not used in this project, but we have shown it because it may be very useful in another iBeacon project, as it gives an estimation of the distance between an iOS device and each iBeacon in range. Ranging iBeacons can be started and stopped by using [_locationManager startRangingBeaconsInRegion:(CLBeaconRegion *)region] and [_locationManager stopRangingBeaconsInRegion:(CLBeaconRegion *)region] respectively.

- monitoringDidFailForRegion: This method tells us whether something is wrong in the monitoring of any region. Never forget to implement it.

```
- (void)locationManager:(CLLocationManager *)manager
didRangeBeacons:(NSArray *)beacons inRegion:(CLBeaconRegion *)
region {

    if ([beacons count] == 0) {
        return;
    }

    CLBeacon *b = beacons[0];

    if (b.proximity == CLProximityFar) {

        NSLog(@"Far");
    }

    if (b.proximity == CLProximityNear) {

        NSLog(@"Near");
    }

    if (b.proximity == CLProximityImmediate) {

        NSLog(@"Immediate");
    }
```

```
        if (b.proximity == CLProximityUnknown) {

            NSLog(@"Unknown");
        }
    }


    - (void)locationManager:(CLLocationManager *)manager monitoringDid
    FailForRegion:(CLRegion *)region withError:(NSError *)error {

        UIAlertView *alert = [[UIAlertView alloc] initWithTitle:NSLoca
    lizedString(@"Error",nil)

    message:[NSString stringWithFormat:@"Region Monitoring
    Failed for the region: %@\n%@", [region identifier],[error
    localizedDescription]]

                                                    delegate:self
                                         cancelButtonTitle:@"Ok"

    otherButtonTitles:nil,nil];

        NSLog(@"%@", [error localizedDescription]);

        [alert show];
    }
```

Writing code for PinsViewController

This view controller manages the PINs that are required to authorize your relatives and friends so that they can access your garage. It works almost in the same way as the ActivationsTableViewController in the Power Plug project. So, we won't spend much time on it.

The main difference here is that we have to enter just a PIN, and it doesn't make sense to create a screen for that. We take advantage of a feature of UIAlertView. By setting its style to `UIAlertViewStylePlainTextInput`, it presents a text field in which we can enter the PIN. This is very easy and convenient.

You should be able to write this view controller yourself and compare your results with the downloaded code. Let's give it a try.

Testing and tuning

We are now ready to test this project and impress our neighbors. First, you have to set the iBeacon parameters. If you are using RedLab iBeacon, you can set it by using the iOS app that is available from the iTunes Store for free (`https://itunes.apple.com/it/app/redbear-beacontool/id828819434?l=en&mt=8`).

You have to enter the following values:

- **UUID**: 00000000-0000-0000-0000-0000000000FF
- **Major**: 0
- **Minor**: 1
- **Advertising Interval**: 250 ms
- **TX Power**: 0

> Double-check the UUID. It's a long string, and any error prevents the iBeacon from being recognized by the application.

Tuning the iBeacon parameters

Once everything works as expected, you can try reducing the TX power and/or Advertising Interval. The lower they are, the more battery you save, and the longer the iBeacon works without you having to replace the batteries. Moreover, reducing the TX power allows you to send the opening command when you are closer to your garage. Let's make some tests to detect the best values for you.

If you are using a different iBeacon, ask the manufacturer how to set the parameters so they are exactly the same.

Now, you should place the iBeacon near your garage door at a high position and leave it turned off.

Wire the relay contacts or the MOSFET pins to your garage door opener (please refer to the electric diagrams provided at the beginning of this chapter). There are a lot of different models out there, so you need to do this yourself. The general advice is that you have to put the relay exit (or the exit connected to the MOSFET) in parallel to the push button that you use to manually open and close the garage door. Take a look at your garage door opener instruction manual for more information and directions.

Before starting the test phase, we need to set up the Arduino code to properly clean up the EEPROM and store the master PIN. To do this, perform the following procedure:

1. In the `setup` function, comment out the following lines, which clean up the EEPROM and store the master PIN:

```
for (int i = 0; i < 6*NUMBER_OF_PINS; i++)
   EEPROM[i] = 0;

// Set the master PIN

EEPROM[0] = 1;      // Don't change this
EEPROM[1] = '1';
EEPROM[2] = '2';
EEPROM[3] = '3';
EEPROM[4] = '4';
EEPROM[5] = '5';
```

 You can change the master PIN (12345) to your preferred code here.

2. Upload the code to Arduino.

3. Comment the previous code again and upload it to Arduino. Now, the EEPROM is cleared and the master PIN is stored.

Now, when you open the app, you will see a message. You have to respond to this message by selecting **Allow**:

In the app, open the Configuration tab, scan for the RF8001, choose your personal PIN, enter it in the PIN field, and then tap on Set House Position. A few seconds later, you should see the longitude and latitude of your house, as acquired by the GPS.

Set the **House Region Size** to 100 meters. Change it to a higher value only if your garage door opens unexpectedly when you are inside your house or if you have a very large house (lucky you!).

Tap the **PINS** tab, enter the master PIN (12345, if you have not changed it in the Arduino code), and tap *Enter*. You should see an empty list. Tap on the add button (**+**) and enter your personal PIN that you chose before.

Tap on the Main tab and then tap on the **Open** button. Now, your garage should open and then close about 30 seconds later.

> To change the closing delay time, you have to change the value of CLOSING_DOOR_INTERVAL in the Arduino code. Moreover, to operate the garage door opener, Arduino shorts the control line for about 300 ms. If this is not enough for your device, you can change the delay in the `pulseOutput` function.

Now, we are going to test the most exciting feature — opening the garage automatically:

1. Open the Main tab again.
2. You should see the **In House Region** indicator turned on, the **In Garage Region** turned off, and the **Ready To Open** turned on.
3. If the **Ready To Open** indicator is off, tap on **Simulate Exit**.
4. Close your app, sending it to the background or closing it from the task list.
5. Turn on the iBeacon.
6. You should see a notification on the screen and hear a short sound, and your garage should start opening.
7. Now, if you move around and inside your house, your garage shouldn't open anymore (the **Ready To Open** indicator should remain off).
8. Drive away from your house until you hear a sound notification from your iOS device. Please drive safely. You don't need to look at your iOS device while driving. The sound notification alerts you.

9. Make a stop and check the application. Now, the **Ready To Open** indicator should be on.

10. Drive back to your house. As soon as you are close enough to the iBeacon, you should hear a notification, and the garage door should start opening.

Now, you are ready to impress your neighbors!

How to go further

Some improvements that could be done to this project are as follows:

*   Setting the automatic door closing interval directly from the app instead of changing the Arduino code.
*   Turning the garage lights on and off on entering and exiting the garage.
*   Opening the garage door from the inside by using the noise generated by the car engine instead of manually. This requires listening to the motor noise from the iOS device and comparing it with a pre-recorded noise of the motor. This is done to avoid a situation where a noise inside or outside the garage unexpectedly opens the door. For signal comparison, you can use an operation called the correlation of two signals (take a look at the Accelerate framework that is available in iOS), but you have to pay attention to the fact that the acquired signal and the pre-recorded signal may have different lengths and/or may be time shifted. Okay, this is a big challenge, but this is the last chapter of the book, and you should be a master of Arduino and iOS programming by now. Digital signal processing is an art that you may be interested in.
*   Checking whether the garage door is effectively closed by using a reed switch, a hall effect sensor, or an ultrasonic distance sensor, and having a notification sent to your phone. You may need the WiFi Shield to take advantage of one of the available IoT services to send the notification.

Summary

While building this project, you learned a lot, especially with regards to iOS. You learned how to create and manage the geofencing and iBeacon regions. This opens the door to many different projects on iOS with or without the Arduino integration. Moreover, you learned how to monitor changes in properties (KVO), which is a technique that lies at the base of good programming together with the help of Model-View-Controller model. This can be applied many times. On Arduino, you learned how to use EEPROM to store information that needs to be permanently stored on the board to control the behavior of your programs.

This project ends this long journey of Arduino and iOS programming and the two platforms' integration.

I hope that you had fun reading the book, coding, and building at least some of the proposed projects (or maybe all!). Mostly, I hope that you learned more about Arduino, iOS, and their integration so that from now on, you can design and build your own projects.

Have a great time making some revolutionary and game-changing projects!

Index

A

Accelerated Framework
about 159
URL 159
accelerometer
defining 107, 108
activation 61
Adafruit Bluefruit LE nRF8001 breakout
URL 13
ADXL345 111
AltBeacon
about 163
URL 163
analog to digital converter (ADC) 22
app
running 45
URL 45
Apple
rules 4
URL 7
Apple Developer Program (ADP) 4
Apple documentation
URL 194
appliances
controlling 60
Arduino
and development environment setup 4, 5
and iOS devices, communication methods
   between 7, 8
IDE installation 5, 6
Arduino and iOS
hardware requirements 2
references 1, 2
software requirements 2

Arduino.cc
versus Arduino.org conflict 3
Arduino code
about 17, 18, 175
additional required libraries, installing 19
Arduino side, testing 27
Arduino side, tuning 27
global variables 20
libraries initialization 20
logic, implementing 22-25
main program 177
references 27
setting up 20, 21
setup code 176
URL 17
Arduino code, Constant Volume Controller
decoder main program 149, 150
decoder setup code 149
library, URL 148
main program 150, 151
references 148
setup code 150
Arduino code, iOS guided rover
about 113
additional libraries, installing 113
main program 116-119
motor control functions 114, 115
setup code 114
URL 113
Arduino code, Wi-Fi power plug
defining 61
main program 62-65
setup code 62
URL 61

Arduino IDE 1.6.4
  URL 5
Arduino MEGA
  about 5
  using 16
Arduino PIN current
  URL 147
Arduino platform
  hardware requirements 2, 3
  software requirements 3
Arduino UNO 5
Arduino UNO R3
  URL 2
atomic attribute
  versus nonatomic attribute 37
Automatic Counting Reference (ARC)
  about 36
  URL 36

B

BLE supporting devices
  URL 8
Bluetooth
  cons 9
  pros 9
  versus TCP/IP 8
Bluetooth 4.0
  URL 8
Bluetooth BLE
  using 1
Bluetooth BLE nRF8001
  URL 2
Bluetooth Breakout board 3
Bluetooth Pet Door Locker project
  Arduino code 17
  Door locker requirements 12
  hardware 12
  improvements 53, 54
  iOS code 28

C

CocoaAsyncSocket
  URL 67
code, for RoverViewController
  defining 127-129

rover by voice commands, driving 141
rover, driving by means of iOS device
    movement 136
rover, testing with manual
    driving 132, 133
used, for controlling rover by iOS
    accelerometer 133-136
used, for controlling rover by voice
    commands 137-141
used, to control rover manually 130-132
components, view controller
  UIDatePicker 82
  UISegmentControl 82
Constant Volume Controller
  about 145
  improvements, implementing 162
  requirements 145

D

datasheet
  URL 111
decibel (dB)
  about 156
  URL 156
DeMorgan's theorem 24
Design Patterns
  URL 31
dtostrf
  about 26
  URL 26
Dynamic Domain Name
    Services (DDNS) 102

E

eBay
  URL 106
external devices
  URL 4

F

flash memory
  about 62
  strings 63

H

hardware, Constant Volume Controller
    about 146
    additional electronic components 146
    Arduino code 148, 149
    electronic circuit 147, 148
    iOS code 151
hardware, iBeacon
    about 170
    additional electronic components 170
    electronic circuit 171-174
hardware, iOS guided rover
    about 106
    accelerometer 107, 108
    accelerometer, mounting 112
    additional electronic components 106
    electronic circuit 108-112
    references 106
    rover, turning 112
hardware, project
    assembly latch 13
    defining 12
    electronic circuit 14-17
    electronic components 12
    required materials 12
    servo motor 13
hardware, Wi-Fi power plug
    additional electronics components 58
    defining 58
    electronic circuit 58-60
H-bridge
    about 109
    URL 109

I

iBeacon
    about 163
    design constraints 166-169
    distance calculation 166
    garage door opener requirements 166-169
    hardware 170
    improvements 208, 209
    references 164
    technical overview 164-166
    URL 2
IDE
    URL 6
IDE version
    using 6
Inter-integrated Circuit (I2C) 54, 55
iOS
    and development environment setup 6
    Xcode installation 6, 7
iOS apps
    developing 28
    URL 28
iOS code
    about 178
    application user interface, designing for
        BLEConnectionViewController 31-38
    application user interface, designing for
        PetDoorLockerViewController 39, 40
    code, writing for BLEConnectionView
        Controller 41-45, 187-190
    code, writing for
        GarageViewController 191-203
    code, writing for
        PetDoorLockerViewController 46-52
    code, writing for PinsViewController 204
    defining 28
    iOS app, testing 53
    testing 205-208
    tuning 205-208
    URL 28
    user interface, designing for
        BLEConnectionView
        Controller 180-182
    user interface, designing for
        GarageViewController 183, 184
    user interface, designing for
        PinsViewController 185, 186
    Xcode project, creating 28-31, 178-180
iOS code, Constant Volume Controller
    about 151
    references 151
    testing 161, 162
    tuning 161, 162

user interface, designing for
    VolumeControllerView
    Controller 153, 154
writing,
    for BLEConnectionViewController 154
writing,
    for VolumeControllerView
    Controller 154-161
Xcode project, creating 151-153
iOS code, iOS guided rover
about 120
code, writing for
    BLEConnectionViewController 127
code, writing for
    RoverViewController 127-129
testing 141
tuning 141
URL 120
Xcode project, creating 120-126
iOS code, Wi-Fi power plug
application user interface, designing for
    ActivationsTableViewController 78-85
application user interface, designing for
    PowerPlugViewController 76-78
application user interface, designing for
    WiFiConnectionViewController 75, 76
class adding, for storing information of each
    activation 73, 74
code, writing for
    ActivationsTableViewController 95-99
code, writing for
    ActivationTableViewController 99
code, writing for AppDelegate 88-94
code, writing for
    PowerPlugViewController 94, 95
code, writing for
    WiFiConnectionViewController 85-87
defining 66
new view controller, adding 68-72
testing 99, 100
tuning 99, 100
URL 66
Xcode project, creating 66

iOS guided rover
defining 105
improving 142
requirements 106
iOS platform
hardware requirements 4
software requirements 4
IR LED
defining 147
IR remote control
URL 146
iTunes Store
URL 205

K

key-value observing 194

M

messages, Arduino
Activations 90
Status 90
methods
centralManagerDidUpdateState 42
didDiscoverPeripheral 42
micro SD card, with FAT16
URL 58
Model View Controller
about 74
URL 74
motor wiring 112

N

Network Time Protocol server
URL 62
nonatomic attribute
reference 37
nRF8001 service
and characteristics 43
nRF8001 wiring 15
NTP server
URL 100

O

Objective-C
  about 28
  URL 28
OpenEars
  URL 122
orientations and rotations, iOS device
  URL 135

P

pitch and roll, iOS device
  URL 135
platforms 5
poolACI 23
Posix time 62
PWM signal
  URL 109

R

readings, accelerometer
  calibrating 142
RFduino
  URL 5

S

Segue identifier 98
sensors 54
sensor types
  analog sensor 54
  comparison, between protocols 55
  defining 54
  digital sensors 54
Serial Peripheral Interface (SPI) 54, 55
services
  discovering 49
snprintf
  about 26
  URL 26
Sparkfun DC Motor Driver TB6612FBG
  URL 106
strong attribute 36

T

TB6612FBG datasheet
  defining 109, 110
TCP/IP
  using 1
  versus Bluetooth 8
TCP/IP communication, Arduino
  Ethernet Shield 7
  Wi-Fi Shield 7
Teensy
  URL 5
TRIAC
  references 60
  URL 60

U

Unarchiver
  URL 5
Unix time
  about 62
  URL 62
UPD
  URL 62
user preferences
  URL 42

V

view controller
  components 82
Virtual Private Network
  URL 102

W

Wi-Fi
  cons 8
  pros 8
Wi-Fi power plug
  about 57
  accessing, from anywhere in world 100
  Arduino code 61

DDNS 102
hardware 58
improvements 103
iOS code 66
port forwarding 101, 102
requirements 58
Wi-Fi shield
  URL 58

X

Xcode
  about 4, 28
  URL 28
Xcode class reference 41
Xcode project
  parameters 151
  references 121

Thank you for buying
Arduino iOS Blueprints

About Packt Publishing

Packt, pronounced 'packed', published its first book, *Mastering phpMyAdmin for Effective MySQL Management*, in April 2004, and subsequently continued to specialize in publishing highly focused books on specific technologies and solutions.

Our books and publications share the experiences of your fellow IT professionals in adapting and customizing today's systems, applications, and frameworks. Our solution-based books give you the knowledge and power to customize the software and technologies you're using to get the job done. Packt books are more specific and less general than the IT books you have seen in the past. Our unique business model allows us to bring you more focused information, giving you more of what you need to know, and less of what you don't.

Packt is a modern yet unique publishing company that focuses on producing quality, cutting-edge books for communities of developers, administrators, and newbies alike. For more information, please visit our website at www.packtpub.com.

About Packt Open Source

In 2010, Packt launched two new brands, Packt Open Source and Packt Enterprise, in order to continue its focus on specialization. This book is part of the Packt Open Source brand, home to books published on software built around open source licenses, and offering information to anybody from advanced developers to budding web designers. The Open Source brand also runs Packt's Open Source Royalty Scheme, by which Packt gives a royalty to each open source project about whose software a book is sold.

Writing for Packt

We welcome all inquiries from people who are interested in authoring. Book proposals should be sent to author@packtpub.com. If your book idea is still at an early stage and you would like to discuss it first before writing a formal book proposal, then please contact us; one of our commissioning editors will get in touch with you.

We're not just looking for published authors; if you have strong technical skills but no writing experience, our experienced editors can help you develop a writing career, or simply get some additional reward for your expertise.

Arduino Development Cookbook

ISBN: 978-1-78398-294-3 Paperback: 246 pages

Over 50 hands-on recipes to quickly build and understand Arduino projects, from the simplest to the most extraordinary

1. Get quick, clear guidance on all the principle aspects of integration with the Arduino.

2. Learn the tools and components needed to build engaging electronics with the Arduino.

3. Make the most of your board through practical tips and tricks.

Arduino Android Blueprints

ISBN: 978-1-78439-038-9 Paperback: 250 pages

Get the best out of Arduino by interfacing it with Android to create engaging interactive projects

1. Learn how to interface with and control Arduino using Android devices.

2. Discover how you can utilize the combined power of Android and Arduino for your own projects.

3. Practical, step-by-step examples to help you unleash the power of Arduino with Android.

Please check **www.PacktPub.com** for information on our titles

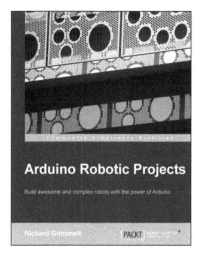

Arduino Robotic Projects

ISBN: 978-1-78398-982-9   Paperback: 240 pages

Build awesome and complex robots with the power of Arduino

1. Develop a series of exciting robots that can sail, go under water, and fly.

2. Simple, easy-to-understand instructions to program Arduino.

3. Effectively control the movements of all types of motors using Arduino.

4. Use sensors, GPS, and a magnetic compass to give your robot direction and make it lifelike.

Arduino Home Automation Projects

ISBN: 978-1-78398-606-4   Paperback: 132 pages

Automate your home using the powerful Arduino platform

1. Interface home automation components with Arduino.

2. Automate your projects to communicate wirelessly using XBee, Bluetooth and WiFi.

3. Build seven exciting, instruction-based home automation projects with Arduino in no time.

Please check **www.PacktPub.com** for information on our titles